D1767654

ADVANCED LINEAR PROGRAMMING: COMPUTATION AND PRACTICE

McGRAW-HILL
INTERNATIONAL
BOOK COMPANY
New York
St Louis
San Francisco
Auckland
Bogotá
Guatemala
Hamburg
Johannesburg
Lisbon
London
Madrid
Mexico
Montreal
New Delhi
Panama
Paris
San Juan
São Paulo
Singapore
Sydney
Tokyo
Toronto

BRUCE A. MURTAGH
Senior Lecturer in Operations Research
University of New South Wales
Australia

Advanced Linear Programming: Computation and Practice

This book was set in Times New Roman by
Eta Services (Typesetters) Ltd., Beccles, Suffolk

British Library Cataloguing in Publication Data

Murtagh, Bruce A
Advanced linear programming
1. Linear programming
I. Title
519.7′2 T57.74 79-42644

ISBN 0-07-044095-6

1 2 3 4 5 MPMP 83 2 1
Printed and bound in the United States of America

To Kim, Andrew and Lisa

CONTENTS

CHAPTER 2
THE REVISED SIMPLEX METHOD

CHAPTER 3
SPARSE MATRIX TECHNIQUES

CHAPTER 4
DUALITY AND POSTOPTIMALITY ANALYSIS

CHAPTER 5
EXTENSIONS TO THE SIMPLEX METHOD

CHAPTER 6
SIMPLEX-BASED NONLINEAR AND INTEGER PROGRAMMING

PART TWO
APPLICATIONS

CHAPTER 7
PROBLEM FORMULATION

CHAPTER 8
LARGE-SCALE MATRIX CONSTRUCTION

CHAPTER 9
COMMERCIAL SYSTEMS: ORGANIZATION OF DATA

CHAPTER 10
COMMERCIAL SYSTEMS: INTERPRETATION OF OUTPUT

PREFACE

There are many textbooks on linear programming (LP) which give an excellent exposition of the mathematics of the subject and of the relevant techniques. However, there is still much that a person needs to learn before solving a real-world problem using one of the many commercially available LP packages. On the other hand, the user manuals which accompany commercial packages are excellent *reference* manuals but not particularly good *teaching* manuals. One objective of this book is to go some way towards bridging this educational gap between theory and practice.

Another objective is to describe some of the advances in LP computational practice that have occurred since the original development of the simplex method by G. B. Dantzig over 30 years ago. As a glance at the Bibliography will indicate, much of this material has hitherto been confined to technical journals and conference proceedings.

In order to keep the book within moderate size one necessarily needs to be selective. I have thus, for example, avoided a lengthy discussion of duality and the dual-simplex method. In general, I have concentrated on aspects relevant to the bulk of current commercial practice, for it is these applications which have provided the impetus to much of the recent research in the field.

So as to give an adequate exposition of model development, it is very useful to consider one example in depth rather than give a brief treatment of several. The application I have chosen is in oil-refinery investment and production planning, not only because this is one of the major commercial applications of LP, but also because it is a convenient vehicle for developing a number of ideas in modeling. Similarly, the matrix-generator language I have used is GAMMA; however, the concepts and approach are equally applicable to other commercially available languages.

In presenting the material I have attempted to strike a level of style which makes it readable, yet still retains a certain amount of rigor. The book is intended for those who will be concerned with developing and running large-scale models in industry, commerce and government. It is suitable for practising analysts and engineers, as well as graduate students who have had a first course in linear algebra and computing. Since this includes most students from most faculties, and the areas of application of LP are so ubiquitous, it is hoped that this book will be useful to a wide spectrum of users.

The material has been developed over a number of years in teaching courses both in university and in industry. Source references are listed in the Bibliography and wherever possible I have cited them at appropriate places within the text.

Any attempt to acknowledge the contribution of colleagues to the existence of this book would be inadequate, but a few people stand out: Professor R. W. H. Sargent, Imperial College, for introducing me to mathematical programming; Norm Helmick, Fluor Corporation, for introducing me to the art of process modeling; and Mike Saunders, Stanford University, not only for a happy and productive association over many years, but also for direct help in carefully reading the manuscript. I would also like to express my sincere appreciation to Chris Paige, McGill University, whose detailed examination of the material and many comments and suggestions gave rise to a substantial improvement in its content. Any errors which have since arisen remain my own.

I would like to thank Letitia Gordon-Wickins for her patient and careful typing, and finally my family, to whom I dedicate this book, for suffering the many lost leisure hours while the book was being written.

PART ONE

COMPUTATION

CHAPTER

ONE

REVIEW OF LINEAR ALGEBRA

In this chapter we shall briefly review a number of pertinent concepts in linear algebra which will be used in subsequent chapters. It is not intended to be an exhaustive review of the subject; there are several excellent texts which cover the subject in great depth. A number of these are listed in the Bibliography (Bodewig, 1959; Dahlquist and Bjorck, 1974; Stewart, 1973; Strang, 1976).

1-1 DEFINITION OF A MATRIX

A matrix is a rectangular array of real numbers. The ijth element a_{ij} of the matrix $\mathbf{A}$ is in the ith row and jth column.

Example

$$\mathbf{A} = \begin{bmatrix} a_{11} & a_{12} & a_{13} & a_{14} \\ a_{21} & a_{22} & a_{23} & a_{24} \\ a_{31} & a_{32} & a_{33} & a_{34} \end{bmatrix} \tag{1-1}$$

A matrix of order $(m \times n)$ has m rows and n columns. A square matrix has $m = n$.

The matrix $\mathbf{A}^\top$ is called the *transpose* of $\mathbf{A}$ if the element a_{ij} in $\mathbf{A}$ is equal to the element a_{ji} in $\mathbf{A}^\top$ for all i and j.

Example

If $$\mathbf{A} = \begin{bmatrix} 1 & 4 & 6 \\ 5 & 7 & 9 \end{bmatrix} \quad \text{then} \quad \mathbf{A}^\top = \begin{bmatrix} 1 & 5 \\ 4 & 7 \\ 6 & 9 \end{bmatrix}$$

A matrix $\mathbf{A}$ is *symmetric* if $\mathbf{A} = \mathbf{A}^\top$.

1-2 DEFINITION OF A VECTOR

Let $x_1, x_2, \ldots, x_n$ be any n real numbers and $\mathbf{x}$ an ordered set of these numbers, i.e.,

$$\mathbf{x} = \begin{bmatrix} x_1 \\ x_2 \\ \vdots \\ x_n \end{bmatrix} \tag{1-2}$$

then $\mathbf{x}$ is called a vector of order n. The ith component of $\mathbf{x}$ is denoted by x_i. Since the numbers are arranged vertically, $\mathbf{x}$ is often referred to as a *column* vector. Note that it could equally be regarded as an $(n \times 1)$ matrix. The set of all such vectors forms a *vector space* E^n of dimension n.

Henceforth we shall denote a *row* vector by

$$\mathbf{x}^\top = (x_1 \quad x_2 \quad \ldots \quad x_n) \tag{1-3}$$

This could equally be regarded as a $(1 \times n)$ matrix. We shall denote the jth column of $\mathbf{A}$ as the vector $\mathbf{a}_j$, and the ith element of this vector as $(\mathbf{a}_j)_i$, or simply a_{ij}. (Both forms have their merits.)

1-3 MATRIX AND VECTOR ARITHMETIC OPERATIONS

Since vectors can be regarded as a special type of matrix, the laws of matrix arithmetic apply equally to these. For compactness we shall use "sigma notation," where, for example, $x_1 + x_2 + \cdots + x_n$ is written as $\sum_{i=1}^{n} x_i$.

1-3-1 Addition

$$\mathbf{C} = \mathbf{A} + \mathbf{B} \tag{1-4}$$

is obtained by adding each element, as long as $\mathbf{A}$ and $\mathbf{B}$ are of the same order $(m \times n)$, i.e.,

$$c_{ij} = a_{ij} + b_{ij} \qquad \text{all } i \text{ and } j.$$

Thus

$$\mathbf{A} + \mathbf{B} = \mathbf{B} + \mathbf{A} \tag{1-5}$$

$$(\mathbf{A} + \mathbf{B})^\top = \mathbf{A}^\top + \mathbf{B}^\top \tag{1-6}$$

1-3-2 Multiplication of Matrices

The product $\mathbf{C} = \mathbf{AB}$ can be achieved if and only if the number of columns in $\mathbf{A}$ equals the number of rows in $\mathbf{B}$.

If $\mathbf{A}$ is of order $(m \times t)$ and $\mathbf{B}$ is of order $(t \times n)$ then $\mathbf{C} = \mathbf{AB}$ is of order

$(m \times n)$ and its elements c_{ij} are given by

$$c_{ij} = \sum_{k=1}^{t} a_{ik} b_{kj} \qquad i = i, \ldots, m, \qquad j = 1, \ldots, n \tag{1-7}$$

Note that $\mathbf{AB} \neq \mathbf{BA}$ in general and $(\mathbf{A} + \mathbf{B})\mathbf{C} = \mathbf{AC} + \mathbf{BC}$. (Whenever $\mathbf{AB}$ is written it will be implicitly assumed that the product is defined.)

As special cases of this rule we observe the *inner product* of two vectors $\mathbf{a}$ and $\mathbf{b}$ is given by

$$\mathbf{a}^\top \mathbf{b} = \sum_{i=1}^{n} a_i b_i = \mathbf{b}^\top \mathbf{a} \tag{1-8}$$

The result is of order (1×1), i.e., a scalar.

The *outer product* of two vectors $\mathbf{a}$ and $\mathbf{b}$ is given by

$$\mathbf{a}\mathbf{b}^\top = \begin{bmatrix} a_1 b_1 & a_1 b_2 & \ldots & a_1 b_n \\ a_2 b_1 & a_2 b_2 & & \\ \vdots & & \ddots & \\ a_n b_1 & a_n b_2 & \ldots & a_n b_n \end{bmatrix} \tag{1-9}$$

Notice that the outer product is a full matrix of order $(n \times n)$. It is called a *rank 1* matrix since each column is a multiple of the same vector $(\mathbf{a})$.

An outer product of two vectors of unequal order exists, but of course an inner product does not.

Using Eq. (1-7) we can write a matrix–vector product as

$$\mathbf{Ax} = \begin{bmatrix} \sum_{j=1}^{n} a_{1j} x_j \\ \sum_{j=1}^{n} a_{2j} x_j \\ \vdots \\ \sum_{j=1}^{n} a_{mj} x_j \end{bmatrix} \tag{1-10}$$

The *norm* of the vector $\mathbf{x}$ is given by

$$\| \mathbf{x} \| = (\mathbf{x}^\top \mathbf{x})^{1/2} = (x_1^2 + x_2^2 + \cdots + x_n^2)^{1/2}$$

and as such is a measure of the magnitude ("length") of the vector.

1-3-3 Multiplication by a Scalar

Multiplication of a matrix (or vector) by a scalar implies that every element is multiplied by the scalar. For example,

$$2\mathbf{A} = \begin{bmatrix} 2a_{11} & 2a_{12} & \dots & 2a_{1n} \\ 2a_{21} & 2a_{22} & & \\ \vdots & & \ddots & \\ 2a_{n1} & & & 2a_{nn} \end{bmatrix} \tag{1-11}$$

1-4 THE IDENTITY MATRIX

The symbol **I** stands for a particular type of square matrix with elements 1 down the diagonal and 0 elsewhere.

Thus
$$\mathbf{I} = \begin{bmatrix} 1 & 0 & 0 & 0 \\ 0 & 1 & 0 & 0 \\ 0 & 0 & 1 & 0 \\ 0 & 0 & 0 & 1 \end{bmatrix} \tag{1-12}$$

Sometimes the order of the matrix is written as a subscript (e.g., $\mathbf{I}_4$ above) but usually the order is understood from the context.

Pre- or post-multiplying any matrix or vector by the identity leaves it unchanged.

A special type of vector is denoted by the symbol $\mathbf{e}_j$; this vector has 1 in position j and zeros elsewhere.

Thus
$$\mathbf{e}_3 = \begin{bmatrix} 0 \\ 0 \\ 1 \\ 0 \end{bmatrix} \tag{1-13}$$

Note that the vector $\mathbf{e}_j$ forms the jth column of **I**. Also, $\mathbf{e}_i^{\mathsf{T}}\mathbf{e}_j = 0$, $j \neq i$, i.e., the vectors are mutually orthogonal.

1-5 THE INVERSE OF A MATRIX

If **A** and **B** are two $(n \times n)$ square matrices such that

$$\mathbf{AB} = \mathbf{I} \tag{1-14}$$

then **B** is called the *inverse* of **A** and is denoted by

$$\mathbf{B} = \mathbf{A}^{-1} \tag{1-15}$$

Note that
$$\mathbf{AA}^{-1} = \mathbf{A}^{-1}\mathbf{A} = \mathbf{I} \tag{1-16}$$

1-6 LINEARLY INDEPENDENT VECTORS

A set of vectors $\mathbf{a}_1, \mathbf{a}_2, \ldots, \mathbf{a}_n$ are said to be *linearly independent* if and only if, for all real α_j,

$$\sum_{j=1}^{n} \alpha_j \mathbf{a}_j = 0 \tag{1-17}$$

implies that $\alpha_j = 0$ for all $j = 1, \ldots, n$.

That is, no one vector $\mathbf{a}_j$ can be written as a linear combination of one or more of the other vectors.

Example The two vectors,

$$\mathbf{a}_1 = \begin{bmatrix} 1 \\ 2 \\ 4 \\ 6 \end{bmatrix} \qquad \mathbf{a}_2 = \begin{bmatrix} 2 \\ 4 \\ 8 \\ 12 \end{bmatrix} \tag{1-18}$$

are **not** linearly independent (and thus are called *linearly dependent*) since $\mathbf{a}_2 = 2\mathbf{a}_1$.

Any set of n linearly independent vectors forms a *basis* in E^n and is said to *span* E^n. This means we can write *any* vector in E^n as a linear combination of the basis vectors. The most easily conceived basis vectors are the vectors $\mathbf{e}_1, \mathbf{e}_2, \ldots, \mathbf{e}_n$, for we can write any vector $\mathbf{x}$ in the form

$$\mathbf{x} = x_1\mathbf{e}_1 + x_2\mathbf{e}_2 + \cdots + x_n\mathbf{e}_n \tag{1-19}$$

This enables us to place a geometric interpretation on a vector, since we can relate $\mathbf{x}$ to a coordinate system if we regard $\mathbf{e}_j$ as the unit vector, along the jth axis. (The reader may wish to draw a vector sum to verify Eq. (1-19).)

A subspace S of E^n *spanned* by a set of vectors is the set of all possible linear combinations of those vectors.

1-7 NONSINGULAR MATRICES

We can also consider the possibility of using the columns of $\mathbf{A}$, $\mathbf{a}_j$, $j = 1, \ldots, n$, as a set of basis vectors. It can readily be established that the maximum number of linearly independent columns of $\mathbf{A}$ is equal to the maximum number of linearly independent rows of $\mathbf{A}$. This number is called the "rank" r of $\mathbf{A}$, and it follows that $r \leqslant \min(m, n)$.

A matrix $\mathbf{A}$ is *nonsingular* if it is square ($m = n$) and of full rank ($r = n$). Only a nonsingular matrix possesses an inverse. In this case the columns of $\mathbf{A}$ span E^n and we can use them as a basis to represent any vector in E^n.

If **A** and **B** are nonsingular matrices then their product is nonsingular, and

$$(\mathbf{AB})^{-1} = \mathbf{B}^{-1}\mathbf{A}^{-1} \tag{1-20}$$

Conversely, note that the product **AB** may be nonsingular even though **A** and **B** may each be singular (since they may be rectangular).

Note the similar result

$$(\mathbf{AB})^{\top} = \mathbf{B}^{\top}\mathbf{A}^{\top} \tag{1-21}$$

If **AB** = **AC** and **A** is nonsingular, then **B** = **C**. If **A** is singular, we cannot make such an assertion.

1-8 MATRIX REPRESENTATION OF LINEAR EQUATIONS

A set of m linear equations in n variables $(x_1, \ldots, x_n)$ can be written in the form

$$\begin{bmatrix} a_{11}x_1 + a_{12}x_2 + \cdots + a_{1n}x_n = b_1 \\ a_{21}x_1 + a_{22}x_2 + \cdots + a_{2n}x_n = b_2 \\ \vdots \qquad \vdots \qquad\qquad \vdots \qquad \vdots \\ a_{m1}x_1 + a_{m2}x_2 + \cdots + a_{mn}x_n = b_m \end{bmatrix} \tag{1-22}$$

or, more conveniently, as the matrix equation

$$\mathbf{Ax} = \mathbf{b} \tag{1-23}$$

A unique solution is obtained if **A** is square ($m = n$) and nonsingular, in which case we have (by premultiplying both sides by $\mathbf{A}^{-1}$)

$$\mathbf{A}^{-1}\mathbf{A}x = \mathbf{I}x = \mathbf{A}^{-1}\mathbf{b}$$

i.e.,

$$\mathbf{x} = \mathbf{A}^{-1}\mathbf{b} \tag{1-24}$$

This is a convenient way of expressing the solution **x**, but there are far better ways of computing the solution than forming $\mathbf{A}^{-1}$ explicitly and premultiplying it into **b**; we shall return to this point in Sec. 1-13.

If $m > n$, the set of equations is said to be *overdetermined* (more equations than unknowns), and if $m < n$ (more unknowns than equations) the set is said to be *underdetermined*. When $m < n$, and rank $(\mathbf{A}) = m$, Eq. (1-23) has $n - m$ degrees of freedom in its solution, in that we can extract m linearly independent columns from **A** and solve for the m associated variables in terms of the remaining $n - m$.

Using our notation of $\mathbf{a}_j$ for the jth column of **A**, we could write the matrix equation (1-23) in the form

$$\mathbf{a}_1x_1 + \mathbf{a}_2x_2 + \mathbf{a}_jx_j + \cdots + \mathbf{a}_nx_n = \mathbf{b} \tag{1-25}$$

Thus, in solving the system of equations, we could regard it as finding the linear combination of the vectors $\mathbf{a}_1, \ldots, \mathbf{a}_n$ which equals the vector **b**. The coefficients of this linear combination are the elements $x_1, x_2, \ldots, x_n$ of the vector **x**. If **A** is nonsingular then the vectors $\mathbf{a}_1, \ldots, \mathbf{a}_n$ are linearly independent, so that there is one and only one such combination.

This ties in with our definition of linearly independent vectors, for if we solve $\mathbf{Ax} = \mathbf{0}$ and $\mathbf{A}$ is nonsingular, the only solution is $\mathbf{x} = \mathbf{0}$. The use of Eq. (1-25) is also a good way of evaluating a matrix–vector product in the computer, if matrices are stored by columns rather than by rows. In using Eq. (1-25) we only have to extract each column of the matrix one at a time.

Example

$$\begin{bmatrix} 3 & 0 & 1 \\ 1 & 2 & 0 \\ 4 & 3 & 5 \end{bmatrix} \begin{bmatrix} 1 \\ 2 \\ 3 \end{bmatrix} = \begin{bmatrix} 3 \\ 1 \\ 4 \end{bmatrix} 1 + \begin{bmatrix} 0 \\ 2 \\ 3 \end{bmatrix} 2 + \begin{bmatrix} 1 \\ 0 \\ 5 \end{bmatrix} 3$$

$$= \begin{bmatrix} 3 \\ 1 \\ 4 \end{bmatrix} + \begin{bmatrix} 0 \\ 4 \\ 6 \end{bmatrix} + \begin{bmatrix} 3 \\ 0 \\ 15 \end{bmatrix} = \begin{bmatrix} 6 \\ 5 \\ 25 \end{bmatrix}$$

The reader may wish to verify that the same result is obtained using the standard formula, Eq. (1-10).

One further notion that Eq. (1-25) reinforces is that each column $\mathbf{a}_j$ of $\mathbf{A}$ is associated with a particular variable, x_j.

1-9 PARTITIONED MATRICES

A matrix (vector) may be *partitioned* into two or more smaller *submatrices* (subvectors). The laws of addition and multiplication act as if these submatrices were themselves elements of the larger matrix.

Example 1 Partition

$$\mathbf{A} = [\mathbf{A}_1 \,|\, \mathbf{A}_2] \tag{1-26}$$

and correspondingly

$$\mathbf{x} = \begin{bmatrix} \mathbf{x}_1 \\ \hline \mathbf{x}_2 \end{bmatrix} \tag{1-27}$$

Then the equation

$$\mathbf{Ax} = \mathbf{b} \tag{1-28}$$

is equivalent to

$$\mathbf{A}_1\mathbf{x}_1 + \mathbf{A}_2\mathbf{x}_2 = \mathbf{b} \tag{1-29}$$

Example 2 Partition

$$\mathbf{A} = \left[\begin{array}{c|c} \mathbf{A}_1 & \mathbf{A}_2 \\ \hline \mathbf{A}_3 & \mathbf{A}_4 \end{array}\right] \tag{1-30}$$

and correspondingly $$\mathbf{x} = \left[\begin{array}{c}\mathbf{x}_1 \\ \hline \mathbf{x}_2\end{array}\right] \tag{1-31}$$

and $$\mathbf{b} = \left[\begin{array}{c}\mathbf{b}_1 \\ \hline \mathbf{b}_2\end{array}\right] \tag{1-32}$$

Then the equation $\mathbf{Ax} = \mathbf{b}$ is equivalent to

$$\begin{aligned} \mathbf{A}_1\mathbf{x}_1 + \mathbf{A}_2\mathbf{x}_2 &= \mathbf{b}_1 \\ \text{and} \qquad \mathbf{A}_3\mathbf{x}_1 + \mathbf{A}_4\mathbf{x}_2 &= \mathbf{b}_2 \end{aligned} \tag{1-33}$$

1-10 ELEMENTARY TRANSFORMATIONS

Modifying a single row or column of a matrix is achieved by a very simple matrix multiplication, called an *elementary transformation*.

Consider a square $n \times n$ matrix $\mathbf{A}$. The jth column of $\mathbf{A}$ can be written as the vector

$$\mathbf{a}_j = \mathbf{Ae}_j \tag{1-34}$$

Similarly, the ith row of $\mathbf{A}$ can be written as the vector $\mathbf{e}_i^\top\mathbf{A}$, and the ijth element of $\mathbf{A}$ can be written as the scalar $\mathbf{e}_i^\top\mathbf{Ae}_j$.

The matrix $\mathbf{Ae}_j\mathbf{e}_j^\top$ is simply an $(n \times n)$ matrix with zeros everywhere except in the jth column

$$\mathbf{Ae}_j\mathbf{e}_j^\top = \begin{bmatrix} & a_{1j} & \\ & \vdots & \\ \mathbf{0} & a_{ij} & \mathbf{0} \\ & \vdots & \\ & a_{nj} & \end{bmatrix} \tag{1-35}$$

Thus $$\mathbf{I} + \mathbf{Ae}_j\mathbf{e}_j^\top = \begin{bmatrix} 1 & & & & & a_{1j} & & & \\ & 1 & & & & & & & \\ & & 1 & & & \vdots & & & \\ & & & 1 & & & & & \\ & & & & 1 & \vdots & & & \\ & & & & & 1 + a_{jj} & & & \\ & & & & & \vdots & 1 & & \\ & & & & & & & 1 & \\ & & & & & a_{nj} & & & 1 \end{bmatrix} \tag{1-36}$$

Such matrices are used to delete and insert particular columns into existing matrices. For example, deleting the jth column of an $(n \times n)$ matrix $\mathbf{B}$ and inserting the vector **a** is accomplished by the operation

$$\begin{aligned}\bar{\mathbf{B}} &= \mathbf{B} - \mathbf{B}\mathbf{e}_j\mathbf{e}_j^{\top} + \mathbf{a}\mathbf{e}_j^{\top} \\ &= \mathbf{B}(\mathbf{I} - \mathbf{e}_j\mathbf{e}_j^{\top} + \mathbf{B}^{-1}\mathbf{a}\mathbf{e}_j^{\top})\end{aligned} \tag{1-37}$$

i.e.,

$$\bar{\mathbf{B}} = \mathbf{B}\begin{bmatrix} 1 & & & & & \\ & 1 & & & & \\ & & 1 & & & \\ & & & 1 & \left[\ \mathbf{B}^{-1}\mathbf{a}\ \right] & \\ & & & & 1 & \\ & & & & & 1 \end{bmatrix} \tag{1-38}$$

since the two entries of 1 cancel each other in the jjth position. We shall be performing this type of operation in later chapters.

1-11 THE SHERMAN–MORRISON MATRIX IDENTITY

If

$$\bar{\mathbf{A}} = \mathbf{A} + \mathbf{a}\mathbf{b}^{\top} \tag{1-39}$$

then

$$\bar{\mathbf{A}}^{-1} = \mathbf{A}^{-1} - \frac{\mathbf{A}^{-1}\mathbf{a}\mathbf{b}^{\top}\mathbf{A}^{-1}}{1 + \mathbf{b}^{\top}\mathbf{A}^{-1}\mathbf{a}} \tag{1-40}$$

Note that the denominator in Eq. (1-40) is just a scalar. The reader may wish to verify the formula by pre- or post-multiplying Eq. (1-40) by Eq. (1-39). If $(1 + \mathbf{b}^{\top}\mathbf{A}^{-1}\mathbf{a}) = 0$ then $\bar{\mathbf{A}}$ is singular. (We shall see later that the use of Eq. (1-40) as a *computational* formula should be discouraged; in fact, inverses are convenient for exposition but not particularly useful computationally.)

We shall occasionally use permutation matrices of the form

$$\begin{aligned}\mathbf{P}_{pq} &= \mathbf{I} + (\mathbf{e}_q - \mathbf{e}_p)\mathbf{e}_p^{\top} + (\mathbf{e}_p - \mathbf{e}_q)\mathbf{e}_q^{\top} \\ &= \mathbf{I} - (\mathbf{e}_p - \mathbf{e}_q)(\mathbf{e}_p^{\top} - \mathbf{e}_q^{\top})\end{aligned}$$

Such a matrix has the effect of interchanging columns p and q of any matrix (or vector) which premultiplies it.

Note from Eq. (1-40) that $P_{pq} = P_{pq}^{-1} = P_{pq}^{\top}$. (Not all permutation matrices are symmetric, but they are all *orthogonal*, i.e., $P^{\top} = P^{-1}$). Premultiplying a matrix (or vector) by P_{pq} has the effect of interchanging the *rows*.

1-12 SPARSE AND DENSE MATRICES

A matrix with most of its elements a_{ij} nonzero is referred to as a *dense* matrix. On the other hand, a matrix with most of its elements zero is referred to as a *sparse* matrix. The *structure* of a matrix is the pattern in which the nonzero coefficients appear. For example, a *block-diagonal* matrix is one with nonzero entries grouped in blocks along the diagonal, as shown in Fig. 1-1.

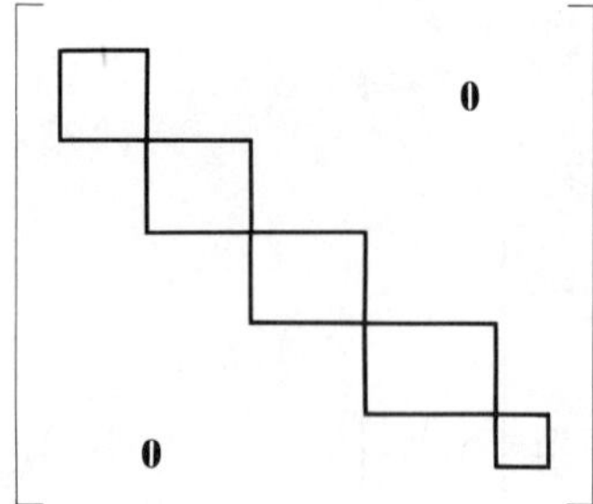

Figure 1-1 A block-diagonal matrix.

Sparse matrixes may well be nonsingular even though they have few nonzero entries; for example, the identity **I** has only one entry per column yet is nonsingular.

A *lower-triangular* matrix **L** is a square matrix with all zeros above the diagonal, and an *upper-triangular* matrix **U** has all zeros below the diagonal. Thus **L** has the structure

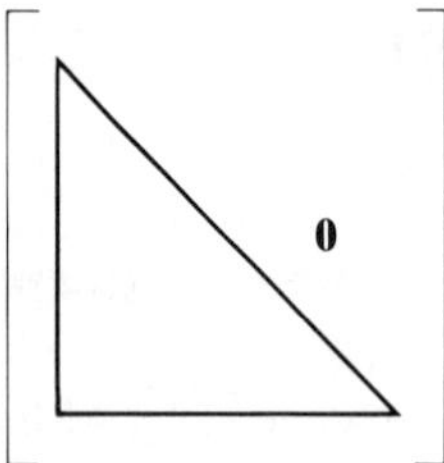

and **U** has the structure

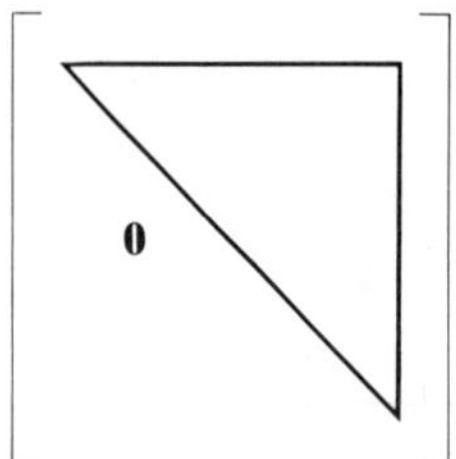

1-13 SOLUTION OF LINEAR EQUATIONS

Consider the matrix equation

$$\mathbf{A}\mathbf{x} = \mathbf{b} \tag{1-41}$$

When **A** is square and nonsingular, it is notationally convenient to represent the solution in the form

$$\mathbf{x} = \mathbf{A}^{-1}\mathbf{b} \tag{1-42}$$

However, it should be emphasized that in solving Eq. (1-41) we do *not* form an explicit representation of $\mathbf{A}^{-1}$ as an $n \times n$ matrix and then postmultiply it by the vector $\mathbf{b}$. The solution $\mathbf{x}$ can be obtained in far fewer steps than obtaining $\mathbf{A}^{-1}$ explicitly. The basic process, called "*gaussian elimination*," is closely related to the simplex method to be discussed in Chapter 2. The procedure is best introduced by an example. Consider the equations

$$
\begin{array}{ll}
(1) & 3x_1 \phantom{{}+2x_2} + x_3 = 6 \\
(2) & x_1 + 2x_2 \phantom{{}+x_3} = 5 \\
(3) & 4x_1 + 3x_2 + 5x_3 = 25
\end{array}
$$

Subtract (1) × 1/3 from (2) and (1) × 4/3 from (3), to give (2)′ and (3)′ respectively,

$$
\begin{array}{ll}
(1) & 3x_1 \phantom{{}+2x_2} + x_3 = 6 \\
(2)' & 2x_2 - \frac{1}{3}x_3 = 3 \\
(3)' & 3x_2 + \frac{11}{3}x_3 = 17
\end{array}
$$

Subtract (2)′ × 3/2 from (3)′, to give (3)″.

We thus have

$$
\begin{array}{ll}
(1) & 3x_1 \phantom{{}+2x_2} + x_3 = 6 \\
(2)' & 2x_2 - \frac{1}{3}x_3 = 3 \\
(3)'' & \frac{25}{6}x_3 = \frac{25}{2}
\end{array}
$$

Back-substitution yields

$$
\begin{aligned}
x_3 &= \tfrac{6}{2} = 3 \\
x_2 &= \tfrac{3}{2} + \tfrac{1}{2} = 2 \\
x_1 &= \tfrac{6}{3} - \tfrac{3}{3} = 1
\end{aligned}
$$

We shall now examine the process by which we obtained the solution.

1-13-1 Elimination

In order to be able to solve for x_1, x_2, x_3 sequentially, we eliminate each variable in turn from successive equations, so that (1), (2), (3) are transformed into (1), (2)′, (3)″.

Consider the nonsingular system

$$
\begin{aligned}
a_{11}x_1 + a_{12}x_2 + \cdots + a_{1n}x_n &= b_1 \\
a_{21}x_1 + a_{22}x_2 + \cdots + a_{2n}x_n &= b_2 \\
\vdots \qquad\quad \vdots \qquad\qquad\quad \vdots \quad &\quad \vdots \\
a_{n1}x_1 + a_{n2}x_2 + \cdots + a_{nn}x_n &= b_n
\end{aligned}
\tag{1-43}
$$

To eliminate x_1 from all equations but the first we subtract from the ith equation $m_{i1} = a_{i1}/a_{11}$ times the first equation for $i = 2, \ldots, n$.

The last $(n-1)$ equations now become

$$\begin{aligned} a_{22}^{(2)}x_2 + \cdots + a_{2n}^{(2)}x_n &= b_2^{(2)} \\ \vdots \qquad\qquad \vdots \quad & \quad \vdots \\ a_{n2}^{(2)}x_2 + \cdots + a_{nn}^{(2)}x_n &= b_n^{(2)} \end{aligned} \tag{1-44}$$

where
$$\begin{aligned} a_{ij}^{(2)} &= a_{ij} - m_{i1}a_{1j} \qquad j = 2, \ldots, n \\ b_i^{(2)} &= b_i - m_{i1}b_1 \qquad i = 2, \ldots, n \end{aligned}$$

This corresponds to the equations (2)′ (3)′ in the example in the preceeding section. We can now ignore x_1 and consider the $(n-1)$ equations in $x_2, \ldots, x_n$.

In a similar manner, to eliminate x_2 from all equations but the second, we subtract from the ith equation $m_{i2} = a_{i2}^{(2)}/a_{22}^{(2)}$ times the second equation, for $i = 3, \ldots, n$.

The last $(n-2)$ equations now become

$$\begin{aligned} a_{33}^{(3)}x_3 + \cdots + a_{3n}^{(3)}x_n &= b_3^{(3)} \\ \vdots \qquad\qquad \vdots \quad & \quad \vdots \\ a_{n3}^{(3)}x_3 + \cdots + a_{nn}^{(3)}x_n &= b_3^{(3)} \end{aligned} \tag{1-45}$$

where $a_{ij}^{(3)} = a_{ij}^{(2)} - m_{i2}a_{2j}^{(2)}$, $j = 3, \ldots, n$, $b_i^{(3)} = b_i^{(2)} - m_{i2}b_2^{(2)}$, $i = 3, \ldots, n$. This corresponds to Eq. (3)″.

In general, we continue the process until after $(n-1)$ steps, when we get

$$\begin{aligned} a_{11}x_1 + a_{12}x_2 + a_{23}x_3 + \cdots + a_{1n}x_n &= b_1 \\ a_{22}^{(2)}x_2 + a_{23}^{(2)}x_3 + \cdots + a_{2n}^{(2)}x_n &= b_2^{(2)} \\ a_{33}^{(3)}x_3 + \cdots + a_{3n}^{(3)}x_n &= b_3^{(3)} \\ \vdots \quad & \quad \vdots \\ a_{nn}^{(n)}x_n &= b_n^{(n)} \end{aligned} \tag{1-46}$$

Applying the formula for step k, $k = 1, \ldots, n-1$,

$$\begin{aligned} m_{ik} &= a_{ik}^{(k)}/a_{kk}^{(k)}, \\ a_{ij}^{(k+1)} &= a_{ij}^{(k)} - m_{ik}a_{kj}^{(k)}, \qquad j = k+1, \ldots, n, \\ b_i^{(k+1)} &= b_i^{(k)} - m_{ik}b_k^{(k)}, \qquad i = k+1, \ldots, n \end{aligned} \tag{1-47}$$

where we subtract from the ith equation m_{ik} times the kth equation ($a_{ij}^{(1)}$ corresponds to the original a_{ij}).

Note that the procedure breaks down if $a_{kk}^{(k)}$ is zero. We can take steps to avoid this as long as **A** is nonsingular. This is discussed in Sec. 1-13-3.

Equation (1-46) is now an upper-triangular system, which we can solve by back-substitution.

1-13-2 Back-substitution

Equations (1), (2)′, (3)″ are an upper-triangular system of the form $\mathbf{Ux} = \mathbf{b}$

$$\begin{aligned}
u_{11}x_1 + u_{12}x_2 + \cdots + u_{1n-1}x_{n-1} + u_{1n}x_n &= b_1 \\
u_{22}x_2 + \cdots + u_{2n-1}x_{n-1} + u_{2n}x_n &= b_2 \\
\vdots \quad & \quad \vdots \\
u_{n-1n-1}x_{n-1} + u_{n-1n}x_n &= b_{n-1} \\
u_{nn}x_n &= b_n
\end{aligned} \tag{1-48}$$

Assuming $u_{ii} \neq 0$, $i = 1, \ldots, n$, we compute $x_1, \ldots, x_n$ backwards from x_n,

$$\begin{aligned}
x_n &= \frac{b_n}{u_{nn}} \\
x_{n-1} &= \frac{b_{n-1} - u_{n-1n}x_n}{u_{n-1n-1}} \\
&\vdots \\
x_i &= \frac{b_i - \sum_{j=i+1}^{n} u_{ij}x_j}{u_{ii}} \qquad i = n-2, \ldots, 1
\end{aligned} \tag{1-49}$$

A lower-triangular system $\mathbf{Lx} = \mathbf{b}$ can be solved sequentially by "forward" substitution in a similar manner.

1-13-3 Row Interchanges

As long as $\mathbf{A}$ is nonsingular we can take steps to avoid the element $a_{kk}^{(k)}$, called the *pivot element*, from being zero (or close to zero) by interchanging the kth row with some other row r, $k < r \leqslant n$, for which $a_{rk}^{(k)}$ is not zero. Such a row must exist, for otherwise the first k columns are linearly dependent, which contradicts our assumption that $\mathbf{A}$ is nonsingular.

The strategy known as *partial pivoting* chooses the row r corresponding to the largest element in magnitude

$$|a_{rk}^{(k)}| = \max_i \{|a_{ik}^{(k)}|\} \qquad i = k, \ldots, n \tag{1-50}$$

Another strategy is to interchange simultaneously both row k with row r and column k with column s, $k < s \leqslant n$, for which $a_{rs}^{(k)}$ is the largest element in magnitude in the entire remaining portion of the matrix. This strategy is called *complete pivoting*.

Such strategies ensure that we avoid small pivot elements, which can give rise to problems of numerical accuracy, but we shall see in Chapter 3 that we may also wish to avoid creation of new nonzeros ("fill-in") in the modified matrix, and this entails a different approach.

1-13-4 LU Factorization

A convenient way of solving the set of equations $\mathbf{Ax} = \mathbf{b}$ in the computer is to factorize $\mathbf{A}$ into the product of a lower-triangular matrix $\mathbf{L}$ and an upper-triangular matrix $\mathbf{U}$,

$$\mathbf{A} = \mathbf{LU}\,. \tag{1-51}$$

Obtaining the solution $\mathbf{x}$ is a two-stage operation. We first solve

$$\mathbf{Ly} = \mathbf{b} \tag{1-52}$$

for $\mathbf{y}$, by forward-substitution; we then solve

$$\mathbf{Ux} = \mathbf{y} \tag{1-53}$$

for $\mathbf{x}$, by back-substitution.

It turns out that we obtain the factors $\mathbf{L}$ and $\mathbf{U}$ during gaussian elimination. Consider the modified matrix at the kth step of the elimination

$$\mathbf{A}^{(k)} = \begin{bmatrix} a_{11} & a_{12} & \cdots & a_{1k} & \cdots & a_{1n} \\ & a_{22}^{(2)} & & a_{2k}^{(2)} & & a_{2n}^{(2)} \\ & & \ddots & \vdots & & \vdots \\ & 0 & & a_{kk}^{(k)} & & a_{kn}^{(k)} \\ & & & \vdots & & \vdots \\ & & & a_{nk}^{(k)} & & a_{nn}^{(k)} \end{bmatrix} \tag{1-54}$$

Consider also the elementary transformation

$$\mathbf{M}_k = \mathbf{I} - \mathbf{m}_k \mathbf{e}_k^{\top} \tag{1-55}$$

where

$$\mathbf{m}_k = \begin{bmatrix} 0 \\ \vdots \\ 0 \\ a_{k+1k}^{(k)}/a_{kk}^{(k)} \\ \vdots \\ a_{ik}^{(k)}/a_{kk}^{(k)} \\ \vdots \\ a_{nk}^{(k)}/a_{kk}^{(k)} \end{bmatrix} \tag{1-56}$$

We have
$$\mathbf{M}_k \mathbf{A}^{(k)} = (\mathbf{I} - \mathbf{m}_k \mathbf{e}_k^{\top})\mathbf{A}^{(k)} = \mathbf{A}^{(k)} - \mathbf{m}_k \mathbf{e}_k^{\top} \mathbf{A}^{(k)} \tag{1-57}$$

$\mathbf{e}_k^\top \mathbf{A}^{(k)}$ is just the kth row of $\mathbf{A}^{(k)}$, so Eq. (1-57) is a compact way of representing the kth step of the elimination process: subtract from the ith equation $m_{ik} = a_{ik}^{(k)}/a_{kk}^{(k)}$ times the kth equation for $i = k + 1, \ldots, n$ (cf. Eq. (1-47)).

Ignoring possible row permutations caused by partial pivoting, we can represent the $(n - 1)$ steps of gaussian elimination in the form

$$\mathbf{M}_{n-1}\mathbf{M}_{n-2} \ldots \mathbf{M}_k \ldots \mathbf{M}_2\mathbf{M}_1\mathbf{A} = \mathbf{U} \tag{1-58}$$

where $\mathbf{U}$ is the upper-triangular matrix in Eq. (1-46).

It is readily apparent from Eq. (1-55) that $\mathbf{M}_k$ is a nonsingular lower-triangular matrix with subdiagonal entries in the kth column only, so the product $\mathbf{M}_{n-1}\mathbf{M}_{n-2} \ldots \mathbf{M}_2\mathbf{M}_1$ is also nonsingular. Using Eq. (1-20), the inverse of the product is given by

$$\mathbf{M}_1^{-1}\mathbf{M}_2^{-1} \ldots \mathbf{M}_k^{-1} \ldots \mathbf{M}_{n-2}^{-1}\mathbf{M}_{n-1}^{-1} \tag{1-59}$$

If $\mathbf{A}$ is nonsingular, then we will have constructed a nonsingular $\mathbf{U}$ corresponding to Eq. (1-46), so we may write $\mathbf{A} = \mathbf{LU}$, where

$$\mathbf{L} = \mathbf{M}_1^{-1}\mathbf{M}_2^{-1} \ldots \mathbf{M}_k^{-1} \ldots \mathbf{M}_{n-2}^{-1}\mathbf{M}_{n-1}^{-1} \tag{1-60}$$

Thus the lower-triangular factor $\mathbf{L}$ is just the product of the inverses of the elementary transformations $\mathbf{M}_k$ used in gaussian elimination. If partial pivoting is used to produce the nonsingular $\mathbf{U}$ from $\mathbf{A}$, then $\mathbf{L}$ and $\mathbf{U}$ are the triangular factors of the permuted $\mathbf{A}$ corresponding to the row interchanges.

If $\mathbf{M}_k = \mathbf{I} - \mathbf{m}_k\mathbf{e}_k^\top$, with $\mathbf{m}_k$ defined as in Eq. (1-56), then

$$\mathbf{M}_k^{-1} = \mathbf{I} + \mathbf{m}_k\mathbf{e}_k^\top \tag{1-61}$$

so

$$\begin{aligned} \mathbf{L} &= (\mathbf{I} + \mathbf{m}_1\mathbf{e}_1^\top)(\mathbf{I} + \mathbf{m}_2\mathbf{e}_2^\top) \ldots (\mathbf{I} + \mathbf{m}_k\mathbf{e}_k^\top) \ldots (\mathbf{I} + \mathbf{m}_{n-2}\mathbf{e}_{n-2}^\top)(\mathbf{I} + \mathbf{m}_{n-1}\mathbf{e}_{n-1}^\top) \\ &= \mathbf{I} + [\mathbf{m}_1, \mathbf{m}_2, \ldots, \mathbf{m}_{n-1}, \mathbf{0}] \\ &= \begin{bmatrix} 1 & & & & \\ m_{21} & 1 & & 0 & \\ m_{31} & m_{32} & \ddots & & \\ \vdots & \vdots & & 1 & \\ m_{n1} & m_{n2} & & m_{nn-1} & 1 \end{bmatrix} \end{aligned} \tag{1-62}$$

Thus, the subdiagonal elements of $\mathbf{L}$ are simply the multipliers, m_{ik}, $i = k + 1, \ldots, n$, used in the elimination process, $k = 1, \ldots, n - 1$.

For solution of equations in the computer there is no need to form $\mathbf{L}$ explicitly as in Eq. (1-62). It is just as convenient to leave it in *product form*, given by the right-hand side of Eq. (1-60), so that forward substitution of Eq. (1-52) is effected by obtaining $\mathbf{y} = \mathbf{L}^{-1}\mathbf{b} = \mathbf{M}_{n-1} \ldots (\mathbf{M}_2(\mathbf{M}_1\mathbf{b}))$, i.e., sequential premultiplication by a single elementary transformation.

CHAPTER

TWO

THE REVISED SIMPLEX METHOD

2-1 STATEMENT OF THE PROBLEM

Using matrix notation, the linear programming (LP) problem may be stated in the form

$$\text{Maximize} \qquad P = \mathbf{c}^\top \mathbf{x} \quad \left(= \sum_{j=1}^{n} c_j x_j \right) \tag{2-1}$$

$$\text{subject to} \qquad \mathbf{A}\mathbf{x} = \mathbf{b} \quad \left(\sum_{j=1}^{n} a_{ij} x_j = b_i, i = 1, \ldots, m \right) \tag{2-2}$$

$$\mathbf{x} \geqslant \mathbf{0} \quad (x_j \geqslant 0, j = 1, \ldots, n) \tag{2-3}$$

P in Eq. (2-1) is called the *objective function*, and Eqs. (2-2) and (2-3) are called *constraints*. $\mathbf{A}$ is an $m \times n$ matrix with $n > m$ (usually n is of order 2–3 times as large as m). We will assume that the equations are linearly independent, that is, we do not have any equation which is a multiple of one or more other equations. This requires that we be careful to avoid expressing the same relationship twice in a different manner, which is most likely to occur when describing the operation of the system under consideration. Inequality constraints may be made into equality constraints as in Eq. (2-2) by the use of so-called "slack variables" which are adjoined to the left-hand side and whose value represents the difference between the original left-hand side and the right-hand side. They are often called *logical variables*; the original variables are called *structural variables* as they have more physical significance.

Example The inequality constraint

$$x_1 + 2.0x_2 + 3.56x_3 \geqslant 7.55$$

becomes, when a slack variable x_4 is used,

$$x_1 + 2.0x_2 + 3.56x_3 - x_4 = 7.55$$

Since x_4 is a slack variable it would usually have a zero coefficient in the cost row vector $\mathbf{c}^\mathsf{T}$.

Since we have many more variables (n) than we have linear equations to solve (m), we can use Eq. (2-2) to solve for m of the variables in terms of the other ($n - m$), as follows:

We partition $\mathbf{A}$ into the submatrices $\mathbf{B}$ and $\mathbf{N}$

$$\mathbf{A} = [\mathbf{B} | \mathbf{N}] \tag{2-4}$$

where $\mathbf{B}$ is a $m \times m$ nonsingular submatrix.

We can thus express Eq. (2-2) in the form

$$\mathbf{B}\mathbf{x}_B + \mathbf{N}\mathbf{x}_N = \mathbf{b} \tag{2-5}$$

where $\mathbf{x}$ has been partitioned into

$$\mathbf{x} = \left[\begin{array}{c} \mathbf{x}_B \\ \hline \mathbf{x}_N \end{array}\right] \tag{2-6}$$

corresponding to the partitioning of $\mathbf{A}$.

Note that each of the variables in $\mathbf{x}$ has a corresponding column in $\mathbf{A}$ associated with it, a fact made obvious if we rewrite Eq. (2-5) with $\mathbf{A}$ partitioned into n columns, as in Eq. (1-25).

Equation (2-5) can be rearranged to the form

$$\mathbf{B}\mathbf{x}_B = \mathbf{b} - \mathbf{N}\mathbf{x}_N \tag{2-7}$$

and since $\mathbf{B}$ is nonsingular,

$$\mathbf{x}_B = \mathbf{B}^{-1}\mathbf{b} - \mathbf{B}^{-1}\mathbf{N}\mathbf{x}_N \tag{2-8}$$

we have thus solved for m of the variables ($\mathbf{x}_B$) in terms of the other ($n - m$) variables ($\mathbf{x}_N$).

$\mathbf{x}_B$ is called the vector of *basic* variables.

$\mathbf{x}_N$ is called the vector of *nonbasic* variables.

2-2 BASIC FEASIBLE SOLUTION

If it were not for the requirement of Eq. (2-3), that $\mathbf{x} \geqslant \mathbf{0}$, the ($n - m$) nonbasic variables could take any arbitrary value. A *basic solution* is defined as one on which the nonbasics are set at their bound, in this case zero.

Since,

$$\mathbf{x}_N = \mathbf{0} \tag{2-9}$$

then, from Eq. (2-8),

$$\mathbf{x}_B = \mathbf{B}^{-1}\mathbf{b} \tag{2-10}$$

since the right-hand term of Eq. (2-8) is zero.

A *basic feasible solution* is therefore one in which all the terms of the vector $\mathbf{B}^{-1}\mathbf{b}$ are nonnegative, for then *all* the terms in the vector $\mathbf{x}$ are nonnegative

$$\mathbf{x} = \left[\frac{\mathbf{x}_B\ (\geqslant \mathbf{0})}{\mathbf{x}_N\ (= \mathbf{0})}\right] \geqslant \mathbf{0}$$

and we have satisfied the second requirement of the problem, Eq. (2-3), for all the variables.

2-3 THE TRANSFORMED PROBLEM

If we partition the vector $\mathbf{c}$ into

$$\mathbf{c} = \left[\frac{\mathbf{c}_B}{\mathbf{c}_N}\right] \tag{2-11}$$

corresponding to the partitioning of $\mathbf{x}$ (Eq. (2-6)), then the objective function, Eq. (2-1), can be expressed in the form

$$P = \mathbf{c}^{\mathsf{T}}\mathbf{x} = [\mathbf{c}_B^{\mathsf{T}} | \mathbf{c}_N^{\mathsf{T}}]\left[\frac{\mathbf{x}_B}{\mathbf{x}_N}\right] \tag{2-12}$$

that is,

$$P = \mathbf{c}_B^{\mathsf{T}}\mathbf{x}_B + \mathbf{c}_N^{\mathsf{T}}\mathbf{x}_N \tag{2-13}$$

But, from Eq. (2-8), we have an expression for $\mathbf{x}_B$ in terms of $\mathbf{x}_N$; thus the objective function is

$$P = \mathbf{c}_B^{\mathsf{T}}(\mathbf{B}^{-1}\mathbf{b} - \mathbf{B}^{-1}\mathbf{N}\mathbf{x}_N) + \mathbf{c}_N^{\mathsf{T}}\mathbf{x}_N \tag{2-14}$$

Rearranging,

$$P = \mathbf{c}_B^{\mathsf{T}}\mathbf{B}^{-1}\mathbf{b} - (\mathbf{c}_B^{\mathsf{T}}\mathbf{B}^{-1}\mathbf{N} - \mathbf{c}_N^{\mathsf{T}})\mathbf{x}_N \tag{2-15}$$

Our original problem, Eqs. (2-1)–(2-3), can thus be expressed in the form

Maximize

$$P = \mathbf{c}_B^{\mathsf{T}}\mathbf{B}^{-1}\mathbf{b} - (\mathbf{c}_B^{\mathsf{T}}\mathbf{B}^{-1}\mathbf{N} - \mathbf{c}_N^{\mathsf{T}})\mathbf{x}_N \tag{2-16}$$

subject to

$$\mathbf{x}_B = \mathbf{B}^{-1}\mathbf{b} - \mathbf{B}^{-1}\mathbf{N}\mathbf{x}_N \qquad (\mathbf{x} \geqslant \mathbf{0}) \tag{2-17}$$

This is called the *transformed problem*. (It is sometimes also referred to as the *current tableau*.)

Writing

$$\boldsymbol{\beta} = \mathbf{B}^{-1}\mathbf{b} \tag{2-18}$$

and

$$\boldsymbol{\pi}^{\mathsf{T}} = \mathbf{c}_B^{\mathsf{T}}\mathbf{B}^{-1} \tag{2-19}$$

the transformed problem can be expressed as

Maximize $$P = \mathbf{c}_B^{\mathsf{T}}\boldsymbol{\beta} - (\boldsymbol{\pi}^{\mathsf{T}}\mathbf{N} - \mathbf{c}_N^{\mathsf{T}})\mathbf{x}_N \tag{2-16a}$$

subject to $$\mathbf{x}_B = \boldsymbol{\beta} - \mathbf{B}^{-1}\mathbf{N}\mathbf{x}_N \tag{2-17a}$$

Since $\mathbf{x}_N = \mathbf{0}$, $\mathbf{x}_B$ takes the numerical value of $\boldsymbol{\beta}$ and the objective function P takes on the numerical value of $\mathbf{c}_B^{\mathsf{T}}\boldsymbol{\beta}$. However, the right-hand terms of the equations are included in the transformed problem even though $\mathbf{x}_N = \mathbf{0}$ since they indicate what happens to P and $\mathbf{x}_B$ as one of the elements of $\mathbf{x}_N$ is increased from zero.

2-4 CONDITIONS FOR OPTIMALITY

The row vector

$$\boldsymbol{\pi}^{\mathsf{T}}\mathbf{N} - \mathbf{c}_N^{\mathsf{T}} \tag{2-20}$$

which premultiplies into $\mathbf{x}_N$ in Eq. (2-16), the *objective row* of the transformed problem, is called the vector of *reduced costs*, since it indicates how much the objective function P increases as $\mathbf{x}_N$ changes.

Let us denote the jth element of the reduced-cost vector, Eq. (2-20), by

$$d_j = z_j - c_j \tag{2-21}$$

where c_j is the jth element of $\mathbf{c}_N^{\mathsf{T}}$

and $$z_j = (\boldsymbol{\pi}^{\mathsf{T}}\mathbf{N})_j = \boldsymbol{\pi}^{\mathsf{T}}\mathbf{a}_j \tag{2-22}$$

where $\mathbf{a}_j$ is the jth column of $\mathbf{N}$, the nonbasic portion of $\mathbf{A}$.

Note that we can evaluate z_j without considering any other column in $\mathbf{N}$: this fact is important for computer implementation, as we shall see.

If the reduced cost d_j of the jth nonbasic variable is positive (or zero), we will not increase P by increasing the value of $(\mathbf{x}_N)_j$ above zero.

Thus, if our transformed problem represents a basic feasible solution with $\mathbf{x}_B = \boldsymbol{\beta} \geqslant \mathbf{0}$ and $\mathbf{x}_N = \mathbf{0}$, then the further condition we require for optimality is that *all* the elements of the reduced-cost vector are positive (or zero). To summarize,

Feasibility $$\beta_i \geqslant 0 \qquad i = 1, \ldots, m \tag{2-23}$$

Optimality $$d_j \geqslant 0 \qquad j = 1, \ldots, n - m \tag{2-24}$$

Example Consider the following transformed problem:

Maximize $$P = 30 - 4x_4 - 5x_5 - 3x_6 - 4x_7$$

subject to
$$x_1 = 5 + 3x_4 - x_5 + 1x_6 - 1x_7$$
$$x_2 = 6 - 7x_4 + 2x_5 - 2x_6 - 2x_7$$
$$x_3 = 7 + 1x_4 - 3x_5 + 3x_6 + 3x_7$$

$$\mathbf{c}_B^{\mathsf{T}}\boldsymbol{\beta} = 30 \qquad \mathbf{x}_N = \begin{bmatrix} x_4 \\ x_5 \\ x_6 \\ x_7 \end{bmatrix}$$

$$\mathbf{x}_B = \begin{bmatrix} x_1 \\ x_2 \\ x_3 \end{bmatrix}$$

$$\boldsymbol{\beta} = \begin{bmatrix} 5 \\ 6 \\ 7 \end{bmatrix} \quad -\mathbf{B}^{-1}\mathbf{N} = \begin{bmatrix} +3 & -1 & +1 & -1 \\ -7 & +2 & -2 & -2 \\ +1 & -3 & +3 & +3 \end{bmatrix}$$

$$[\boldsymbol{\pi}^{\mathsf{T}}\mathbf{N} - \mathbf{c}_N^{\mathsf{T}}] = [4 \quad 5 \quad 3 \quad 4]$$

The transformed problem represents a feasible and optimal solution, since we cannot increase P by increasing x_4, x_5, x_6, or x_7 above zero.

There is only one basis corresponding to a given basic feasible solution such that $\mathbf{B}^{-1}\mathbf{b} > \mathbf{0}$, but if $\boldsymbol{\beta} = \mathbf{B}^{-1}\mathbf{b} \geqslant \mathbf{0}$ has a zero element $\beta_i = 0$ (a *degenerate* basic feasible solution) then in $\mathbf{B}\boldsymbol{\beta} = \mathbf{b}$ we can replace the ith column of $\mathbf{B}$ by any nonbasic column of $\mathbf{A}$ which leaves the new $\mathbf{B}$, say $\bar{\mathbf{B}}$, nonsingular and still $\boldsymbol{\beta} = \bar{\mathbf{B}}^{-1}\mathbf{b} = \mathbf{B}^{-1}\mathbf{b} \geqslant \mathbf{0}$, so there may be several bases corresponding to a degenerate basic feasible solution. This affects the wording of the following theorem:

Theorem For a basic feasible solution to be optimal it is necessary and sufficient that there is a corresponding basis for which $d_j \geqslant 0, j = 1, \ldots, n - m$.

The proof of this theorem is constructive, in the sense that it indicates how the steps of the simplex method proceed.

SUFFICIENCY Since Eq. (2-15) holds for every solution of Eq. (2-8), the maximum of P under the condition $\mathbf{x}_N \geqslant \mathbf{0}$ is obtained for $\mathbf{x}_N = \mathbf{0}$, since all $d_j \geqslant 0$, $j = 1, \ldots, n - m$. Thus if we have a basic feasible solution for which $d_j \geqslant 0$, $j = 1, \ldots, n - m$, then it is optimal.

NECESSITY For a basic feasible solution to be optimal it is necessary that $d_j \geqslant 0$, $j = 1, \ldots, n - m$, otherwise

(*a*) If $d_j < 0$ for some $j = 1, \ldots, n - m$, for which $\boldsymbol{\alpha}_j = \mathbf{B}^{-1}\mathbf{a}_j \leqslant \mathbf{0}$, then the corresponding nonbasic variable $(\mathbf{x}_N)_j$ can increase to any positive value, and Eqs. (2-8) and (2-15) indicate respectively that $\mathbf{x}_B$ remains feasible (nonnegative) and P also increases, tending to infinity as $(\mathbf{x}_N)_j$ does.

(*b*) If $d_j < 0$ for some $j = 1, \ldots, n - m$, for which $(\boldsymbol{\alpha}_j)_i > 0$ for at least one $i = 1, \ldots, m$, then we can proceed to a new basic feasible solution in which $(\mathbf{x}_N)_j$ attains the value

$$\left(\frac{\boldsymbol{\beta}}{\boldsymbol{\alpha}_j}\right)_p = \min_{i \mid (\alpha_j)_i > 0} \left\{\left(\frac{\boldsymbol{\beta}}{\boldsymbol{\alpha}_j}\right)_i\right\}$$

and the corresponding $(\mathbf{x}_B)_p$ is zero. This operation is dealt with in detail in the next section, but suffice it to say that the other basic variables remain feasible, so we have a new solution with an improved value of P, which contradicts our assumption that the previous solution was optimal.

The degenerate case can give rise to the possibility of *cycling* in a set of optimal bases,† whereby a number of column-replacement operations (with strictly negative d_j's) eventually arrive at an earlier basis. There are perturbation procedures to overcome this difficulty so that the proof of necessity goes through. These involve imposing a slight perturbation on the constraints so that degeneracies are removed; a detailed description is given in Dantzig (1963).

Until recently, degeneracies were regarded as being of largely theoretical interest as the incidence of cycling in practice was considered to be low. However, it does appear to be on the increase as problems get larger. The integer programming techniques discussed in Chapter 6 often give rise to massive degeneracies in the derived LP problems (Tomlin, 1970).

2-5 PIVOTING

Let us now consider in detail the proof of necessity in the previous theorem. Assume we have at least onc element of the vector of reduced costs which is negative, say d_q. We consider the possibility of increasing only *one* nonbasic variable from zero, while all the others stay at zero. If more than one reduced cost is negative, then we have an arbitrary choice. It is customary to consider the nonbasic variable with the largest (most negative) reduced cost, since this will increase the value of the objective the most per unit change. There are more elaborate methods of resolving this choice, and these will be discussed in Chapter 3.

Writing the transformed problem out column-by-column, it takes the form

$$\text{Maximize} \quad P = \mathbf{c}_B^\top \boldsymbol{\beta} - d_1(\mathbf{x}_N)_1 - \cdots - d_q(\mathbf{x}_N)_q - \cdots - d_{n-m}(\mathbf{x}_N)_{n-m} \tag{2-25}$$

$$\text{subject to} \quad \mathbf{x}_B = \boldsymbol{\beta} - \boldsymbol{\alpha}_1(\mathbf{x}_N)_1 - \cdots - \boldsymbol{\alpha}_q(\mathbf{x}_N)_q - \cdots - \boldsymbol{\alpha}_{n-m}(\mathbf{x}_N)_{n-m} \tag{2-26}$$

$$\text{where} \quad \boldsymbol{\alpha}_j = \mathbf{B}^{-1}\mathbf{a}_j, \qquad j = 1, \ldots, n - m \tag{2-27}$$

As $(\mathbf{x}_N)_q$ increases from zero, the vector of basic variables is affected, according to the relationship

$$\mathbf{x}_B = \boldsymbol{\beta} - \boldsymbol{\alpha}_q(\mathbf{x}_N)_q \tag{2-28}$$

since all the other nonbasic variables remain at zero.

† Since we are maximizing a concave (linear) function subject to convex (linear) constraints, any local optimum is also a global optimum.

As the numerical value of the scalar $(\mathbf{x}_N)_q$ increases, the vector $\mathbf{x}_B$ changes, and it can be seen from Eq. (2-28) that if any element of the vector $\boldsymbol{\alpha}_q$ is positive, then the corresponding element of $\mathbf{x}_B$ will decrease, and eventually pass through zero for a large enough value of $(\mathbf{x}_N)_q$. It must not go below zero of course, for this would ruin the feasibility of $\mathbf{x}$, and this provides a criterion for setting a value of $(\mathbf{x}_N)_q$.

Specifically,

$$(\mathbf{x}_N)_q = \min_{i \mid \alpha_{iq} > 0} \left\{ \frac{\beta_i}{\alpha_{iq}} \right\} \tag{2-29}$$

where α_{iq} and β_i are the ith elements of $\boldsymbol{\alpha}_q$ and $\boldsymbol{\beta}$ respectively.

Suppose the minimum of Eq. (2-29) occurs for $i = p$, say, then when $(\mathbf{x}_N)_q = \beta_p/\alpha_{pq}$ we have $(\mathbf{x}_B)_p = 0$ and any further increase in $(\mathbf{x}_N)_q$ would make $(\mathbf{x}_B)_p$ negative. Note that there is *no* arbitrary choice in the selection of p; having chosen the nonbasic variable (q) to increase, the basic variable (p) which hits zero first is determined by the values of $\boldsymbol{\alpha}_q$ and $\boldsymbol{\beta}$. (If two or more basic variables hit zero simultaneously, we have a degenerate case; however, we must pivot on one of them.)

We are now in a position where a nonbasic variable has a positive value, and a basic variable is zero. Recall that in the partitioned equation (2-7), $(\mathbf{x}_B)_p$ corresponded to the pth column of $\mathbf{B}$. (It can also be regarded as corresponding to the pth *row* of the tableau equation (2-8), but it suits us here to consider the partitioned form in which each element of $\mathbf{x}$ corresponds to a column of $\mathbf{A}$.) The "pivot" operation is completed when the tableau is rearranged so that $(\mathbf{x}_N)_q$ is basic (with a value β_p/α_{pq}) and $(\mathbf{x}_B)_p$ is nonbasic (at zero). This means redefining the partition of $\mathbf{A}$ so that $\mathbf{a}_q$ replaces the pth column of $\mathbf{B}$, corresponding to $(\mathbf{x}_N)_q$ "entering" the basis and $(\mathbf{x}_B)_p$ "leaving" the basis.

The new $\mathbf{B}$ can be represented in the form of Eq. (1-38), where it will be evident that it remains nonsingular since $\alpha_{pq} = (\mathbf{B}^{-1}\mathbf{a}_q)_p$ is nonzero.

The situation described in part (a) of the necessity proof arises when *no* element of $\boldsymbol{\alpha}_q$ is positive. It can be seen from Eq. (2-28) that $(\mathbf{x}_N)_q$ can then be increased arbitrarily and $\mathbf{x}_B$ would still remain positive. (Recall our comment in Sec. 2-2 that $\mathbf{x}_N$ could be arbitrarily chosen if it were not for the requirement that $\mathbf{x} \geqslant \mathbf{0}$.) Since we chose to increase $(\mathbf{x}_N)_q$ because that resulted in an increased value of our objective function (it had a negative reduced cost), then our optimal solution lies out at infinity, and is called an *unbounded solution*. In practice, obtaining an unbounded solution usually means we have failed to consider an important constraint; an easy thing to do during the first few passes at the problem.

2-6 STEPS OF THE REVISED SIMPLEX METHOD

The above results form the building blocks of the simplex method of solution. There are a number of variants of this method, one of which, the revised simplex algorithm, is the one mostly used in computer codes. Since all but trivial LP problems are solved with a computer, we will discuss the operations required from the point of view of computer implementation of the algorithm. Thus, for example, in the revised simplex method all the elements $-\mathbf{B}^{-1}\mathbf{N}$ of the transformed problem are *never* stored or computed, for not only would it take too much work, but also they are not all needed anyway. (Recall that in the previous section we worked with $\boldsymbol{\beta} = \mathbf{B}^{-1}\mathbf{b}$ and $\boldsymbol{\alpha}_q = \mathbf{B}^{-1}\mathbf{a}_q$; these are the only two columns of the transformed problem that we need.)

Assume that we have a basic feasible solution, with some representation of $\mathbf{B}^{-1}$ and current right-hand side $\boldsymbol{\beta} = \mathbf{B}^{-1}\mathbf{b}$. The steps are as follows:

Step 1. Produce a pricing vector Evaluate

$$\pi^{\mathsf{T}} = \mathbf{c}_B^{\mathsf{T}}\mathbf{B}^{-1} \qquad \text{(i.e., solve } \mathbf{B}^{\mathsf{T}}\pi = \mathbf{c}_B \text{ for } \pi). \tag{2-30}$$

This operation is called backwards transformation (BTRAN) since it looks like postmultiplying the row vector $\mathbf{c}_B^{\mathsf{T}}$ by some representation of $\mathbf{B}^{-1}$.

Step 2. Price out the nonbasic columns Evaluate

$$z_j = \pi^{\mathsf{T}}\mathbf{a}_j \tag{2-31}$$

and subsequently the reduced cost

$$d_j = z_j - c_j \tag{2-32}$$

for each nonbasic column.

Most of the work in this step is in taking each of the columns of $\mathbf{N}$ in turn, and evaluating the inner product $\pi^{\mathsf{T}}\mathbf{a}_j$; obviously quite time-consuming if there are a large number of columns.

Step 3. Select the entering nonbasic variable Choose the most negative reduced cost d_j.

(*a*) If none are less than zero (in practice some small error tolerance), then STOP; we know from our optimality condition that we have an optimal feasible solution.

(*b*) Otherwise, let the most negative reduced cost correspond to $j = q$, say. (Ties may be broken arbitrarily.)

Step 4. Update the entering column Evaluate

$$\boldsymbol{\alpha}_q = \mathbf{B}^{-1}\mathbf{a}_q \qquad \text{(i.e., solve } \mathbf{B}\boldsymbol{\alpha}_q = \mathbf{a}_q \text{ for } \boldsymbol{\alpha}_q). \tag{2-33}$$

This operation is called forward transformation (FTRAN) since it looks like premultiplying the vector $\mathbf{a}_q$ by some representation of $\mathbf{B}^{-1}$.

Step 5. Find the leaving basic variable Select

$$\frac{\beta_p}{\alpha_{pq}} = \min_{i|\alpha_{iq}>0} \left\{\frac{\beta_i}{\alpha_{iq}}\right\} \tag{2-34}$$

(*a*) If none exist, STOP; we have an unbounded solution.
(*b*) Otherwise, let the minimum correspond to $i = p$, say. (Ties may also be broken arbitrarily, but one would usually choose the largest pivot α_{pq}, for reasons to be made clear in Chapter 3.)

Step 6. Pivot $(\mathbf{x}_B)_p$ has a numerical value of zero, and $(\mathbf{x}_N)_q$ has a numerical value of β_p/α_{pq}. $\mathbf{a}_q$ replaces the pth column of $\mathbf{B}$, which is most easily represented as $\mathbf{Be}_p$.

Thus we have

$$\bar{\mathbf{B}} = \mathbf{B} + (\mathbf{a}_q - \mathbf{Be}_p)\mathbf{e}_p^{\top} \tag{2-35}$$

which effectively puts $\mathbf{a}_q$ in and takes $\mathbf{Be}_p$ out of the pth column position.

Thus we have, since $\boldsymbol{\alpha}_q = \mathbf{B}^{-1}\mathbf{a}_q$,

$$\bar{\mathbf{B}} = \mathbf{B}[\mathbf{e}_1, \ldots, \mathbf{e}_{p-1}, \boldsymbol{\alpha}_q, \mathbf{e}_{p+1}, \ldots, \mathbf{e}_m] = \mathbf{BF} \qquad \text{say} \tag{2-36}$$

so

$$\bar{\mathbf{B}}^{-1} = \mathbf{F}^{-1}\mathbf{B}^{-1}, \bar{\boldsymbol{\beta}} = \mathbf{F}^{-1}\boldsymbol{\beta} \tag{2-37}$$

and

$$\mathbf{F}^{-1} = [\mathbf{e}_1, \ldots, \mathbf{e}_{p-1}, \boldsymbol{\eta}_q, \mathbf{e}_{p+1}, \ldots, \mathbf{e}_m] = \mathbf{E} \qquad \text{say} \tag{2-38}$$

where

$$\boldsymbol{\eta}_q = \begin{bmatrix} -\alpha_{iq}/\alpha_{pq}, i \neq p \\ 1/\alpha_{pq}, i = p \end{bmatrix} \tag{2-39}$$

Note that $\mathbf{E}$ has the form

$$\mathbf{E} = \begin{bmatrix} 1 & & & & & & \\ & 1 & & & & & \\ & & 1 & & & & \\ & & & 1 & \boldsymbol{\eta}_q & & \\ & & & & & 1 & \\ & & & & & & 1 \end{bmatrix} \tag{2-40}$$

Thus the only significant information which is required to construct the matrix $\mathbf{E}$ is the vector $\boldsymbol{\eta}_q$ and its column position, p. Better yet, $\mathbf{E}$ is known by storing $\boldsymbol{\alpha}_q$ (not $\boldsymbol{\eta}_q$) since this saves all the divisions by α_{pq} (cf. Eq. 2-39); we can execute just one division by α_{pq} whenever we use $\mathbf{E}$.

The iteration is now complete, and if we did not terminate on either of the stopping criteria in Steps 3 and 5, the process is repeated from Step 1.

In practice, the number of iterations is of order 2–4 times the number of rows, although pursuit of the mathematical theory would give an upper bound to the number of iterations far in excess of that estimate. The total running time has been found to correspond roughly to the equation (Wolfe, 1969; Beale, 1971)

$$\text{Time} = K \times (\text{No. of rows})^3 \tag{2-41}$$

although the constant of proportionality K varies with the type of application.

2-7 FIRST FEASIBLE SOLUTION

We have so far assumed that we have available a basic feasible solution. A separate procedure is required to obtain feasibility, and a number of methods are available for doing this.

The so-called "big-M" method consists of adjoining an "artificial" variable to each of the rows of $\mathbf{A}$ in much the same manner as slack variables, except they have coefficients $-M$ (where M is a large positive number) in the objective function. Thus if the problem is feasible, the optimal solution will have each of these artificial variables nonbasic ($=0$) because of their high cost. However, it provides us with an easy starting basis $\mathbf{B} = \mathbf{I}$, corresponding to all the artificial variables since there is a unique, $+1.0$ coefficient in each of the rows they have been added to. Thus, as long as the right-hand side vector $\mathbf{b}$ has nonnegative elements (which can easily be achieved by multiplying the entire row by -1 if necessary) our starting solution $\mathbf{x}_B = \mathbf{B}^{-1}\mathbf{b}$ is feasible since $\mathbf{B}^{-1} = \mathbf{I}$. This method is simple and does not require any special procedure, the simplex iterations remaining as described in Sec. 2-6. Unfortunately, however, the use of large numbers gives rise to severe problems in numerical accuracy, so the big-M method deserves little credibility.

The approach used in commercial codes is to organize the computations in two distinct phases: Phase I entails obtaining a basic feasible solution and Phase II entails solving the problem from this first feasible solution as described in Sec. 2-6. A simplified Phase I procedure consists of the following three steps:

1. Form as much of the identity $\mathbf{I}$ as possible from the existing structural and slack variables.
2. Fill in the remaining portion of $\mathbf{I}$ with artificial variables.
3. Minimize the sum of infeasibilities = the sum of artificial variables.

Thus the objective function for Phase I is simply the sum of the values of the artificial variables, and the problem is feasible when they are all zero. This is equivalent to assigning a *cost* of 1.0 to each of the artificials; it is possible to assign different costs (or *weights*) to each artificial in order to speed convergence.

If the objective function remains positive at the optimal solution of the Phase I problem, then no feasible solution to the original problem exists.

If the Phase I procedure terminates with all artificial variables nonbasic at zero, then these may be removed from further consideration in Phase II. The constraints are not redundant.

If the Phase I procedure terminates with all artificial variables zero, but at least one artificial variable basic, then we also have a feasible solution to the original problem, but the possibility of redundancy arises. If, further, a row i corresponding to a basic artificial has a nonzero coefficient α_{ij} for a nonbasic j (corresponding to a structural or slack variable), then we may pivot on the nonbasic j and remove the artificial from the basis. If this process removes all of the artificials from the basis, then we have a (degenerate) feasible solution and there are no redundancies. If, on the other hand, this does not remove all of them, each row i corresponding to a basic artificial with zero coefficients α_{ij} for all j corresponds to a redundant (but consistent) constraint, i.e., it is a multiple of one or more other constraints.

If Phase I terminates with positive basic artificials, and application of the above process fails to give an (infeasible) solution with only structurals or slacks in the basis, then we have at least one row i corresponding to a basic artificial with zero coefficients α_{ij} for all nonbasics, indicating inconsistent constraints, e.g., $x_1 + x_2 \leqslant 5$ and $x_1 + x_2 \geqslant 6$.

As a simple example, consider the two equality constraints

$$
\begin{aligned}
1.0x_1 + 2.0x_2 + 3.0x_3 &= 3.0 \\
2.0x_1 + 4.0x_2 + 6.0x_3 &= 6.0
\end{aligned}
$$

They are obviously redundant (but consistent), and adding artificial variables x_4 and x_5 makes them linearly independent

$$
\begin{aligned}
1.0x_1 + 2.0x_2 + 3.0x_3 + x_4 \qquad &= 3.0 \\
2.0x_1 + 4.0x_2 + 6.0x_3 \qquad + x_5 &= 6.0
\end{aligned}
$$

If x_4 is nonbasic at zero, then x_5 will be zero and it will be basic (otherwise the basis $\mathbf{B}$ would be singular).

It is somewhat pedantic (computationally) to distinguish between artificial variables and slack variables. If we consider each row as having a slack variable attached to it, then the matrix is considered to be of the form $[\mathbf{A} \mid \mathbf{I}]$, where the identity corresponds to the slacks. It will be seen in later chapters that slacks may have both upper and lower bounds, as may the structural variables. An *artificial* variable is then simply a slack with an upper (and lower) bound of zero. When both upper and lower bounds are present, a variable may be nonbasic at either its upper bound or its lower bound, and Phase I entails bringing all variables (including structurals) within their bounds.

The Phase I objective row (to be maximized) would consist of the following terms for all variables, including slacks:

$$
\begin{aligned}
c_j &= -1 \qquad \text{if } x_j \text{ is above its upper bound} \\
c_j &= +1 \qquad \text{if } x_j \text{ is below its lower bound} \\
c_j &= 0 \qquad \text{otherwise}
\end{aligned}
$$

These coefficients would be used in producing the pricing vector $\boldsymbol{\pi}^{\mathrm{T}}$ in Step 1 of the simplex method, Eq. (2-30), and in pricing out the nonbasic columns in Step

2, Eq. (2-32). It is also possible to include some weighting W of the original objective row.

Having chosen the entering nonbasic variable, say $(\mathbf{x}_N)_q$, in choosing the leaving basic variable, Step 5, we are not restricted to stopping as soon as one of the infeasible basic variables becomes feasible; we may continue, as more than one may become feasible, and stop as soon as a feasible variable goes infeasible. Rarick (1975) has suggested further that we allow a feasible variable to go infeasible, and only stop when the *sum* of infeasibilities stops decreasing.

The technique known as *goal programming* (see, for example, Salkin and Kornbluth, 1973) essentially applies the Phase I procedure to a set of constraints which reflect a number of multiple objectives (goals) which must be satisfied. Thus the problem is cast in the form

$$\text{Minimize} \quad w_1(x_1^- + x_1^+) + \cdots + w_i(x_i^- + x_i^+) + \cdots + w_m(x_m^- + x_m^+)$$

$$\text{subject to} \quad \begin{array}{llllll} a_{11}x_1 + \cdots + a_{1n}x_n & + x_1^- - x_1^+ & & & = b_1 \\ \vdots \qquad\quad \vdots & \ddots & & & \vdots \\ a_{i1}x_1 + \cdots + a_{in}x_n & & + x_i^- - x_i^+ & & = b_i \\ \vdots \qquad\quad \vdots & & \ddots & & \vdots \\ a_{m1}x_1 + \cdots + a_{mn}x_n & & & + x_m^- - x_m^+ & = b_m \end{array}$$

$$x_j \geqslant 0, j = 1, \ldots, n;\ x_i^+ \geqslant 0, x_i^- \geqslant 0, i = 1, \ldots, m$$

x_i^- and x_i^+ are respectively the shortfall and excess of the sum $\sum_{j=1}^{n} a_{ij}x_j$ against the goal b_i, $i = 1, \ldots, m$, and it can be seen that they are very similar to artificial variables. There may also be other constraints in addition to the goals. The weights $w_i \geqslant 0$, $i = 1, \ldots, m$ reflect the relative importance of the goals and it is obvious that, at the solution, one or other of each set $\{x_i^+, x_i^-\}$ will be nonbasic at zero.

Since it is assumed that not all of the variables x_i^+, x_i^-, $i = 1, \ldots, m$, will be zero at the solution (otherwise the usual Phase I procedure would suffice) the choice of weights w_i becomes crucial. It is possible to devise a "pre-emptive" order in which one goal i is satisfied (x_i^+, x_i^- are both made nonbasic) before any other is considered; this involves solving a sequence of problems in which each goal is considered in turn, and once satisfied is kept as a strict equality constraint for the remaining problems.

CHAPTER

THREE

SPARSE MATRIX TECHNIQUES

3-1 INTRODUCTION

The algebra of the revised simplex method was presented in the previous chapter, and it can be seen that in mathematical terms alone it is a very elegant algorithm. However, it could be argued that much of the success of the algorithm in solving large real-world problems is due to efficient methods of *implementing* the algorithm on digital computers.

It has already been noted in Sec. 2-6 that we do not need to store or operate on the whole of the transformed problem; but it was indicated that we need to have available the columns of **N** to carry out the pricing operation (Step 2) as well as some representation of $\mathbf{B}^{-1}$ (for Steps 1 and 4). Now if the representation of $\mathbf{B}^{-1}$ was as an $m \times m$ dense matrix and **N** was stored as a dense matrix, then the practical limit to problem size would be around $m = 300$, since the storage of $\mathbf{B}^{-1}$ would require around 100,000 words, thus using up most of the central memory in the computer and leaving little room for working with **N**. However, problems with thousands of rows have been solved in practice for many years.

This chapter discusses in detail how the algorithm is implemented, and most of this discussion is centered on two considerations which arise in practice:

1. *Storage.* The matrix **A** is sparse. The density rarely exceeds 5 percent and is usually less than 1 percent.
2. *Rounding error.* Finite precision arithmetic is used, thus limiting the accuracy with which a calculated number is known.

We can solve large problems by using efficient methods of storing **A**, and devising a representation of $\mathbf{B}^{-1}$ which does not increase the density too much. (The explicit inverse of a sparse matrix is not necessarily sparse itself).

Rounding error will accumulate as we proceed with the iterations of the algorithm. We must endeavour to keep it as low as possible, monitor its effect, and clean up our numbers when it becomes appreciable.

3-2 STORAGE

Sparsity arises quite naturally in most applications, usually the larger the matrix the more sparse it is. One rule-of-thumb guide is that most columns will have around 5–10 nonzero entries each. This is because description of a large system usually involves many variables, but each individual variable is subject to only a few constraints. An easily conceived engineering example is a network in which a coefficient a_{ij} will be nonzero only if there is a link from node i to node j of the network. Similarly, description of large government or business organizations requires many variables, but these can be subclassified into groups or departments, each subject to mainly their own organizational constraints.

Because the size of central memory is limited the large sparse **A** matrix is usually stored out of central memory column-by-column as one long sequence of words. However, since only the nonzero coefficients are stored, we need to know not only the numerical value of the coefficient a_{ij} but also its row and column position i and j. The row and column indices are integer numbers, so several of them can be packed into a single word.

For example, integers i and j of, say, 3 and 2 may be represented by the single number 3,002 stored as one word. Thus to extract the indices we divide by 1,000 (using integer arithmetic) and examine the quotient and remainder each time.

Commercial LP codes use a more elaborate means of coding the bit-pattern of the indices to keep the extra computing time to pack and unpack the word as small as possible: they essentially mask off the portion of the word not required.

Various schemes have been adopted for storing the matrix, although they all store **A** column by column, recording the nonzero row entries a_{ij} and the corresponding row index i. One of the more commonly used ones is the *block index scheme* as illustrated in Table 3-1.

Table 3-1 Matrix storage scheme

	Block index
Word 1	Column header: No. of row elements
Word 2	Coeff. 1 Coeff. 2
Word 3	Coeff. 3 Coeff. 4
Word 4	Row 1 Row 2 Row 3 Row 4

In the block index scheme, two numbers a_{ij} are stored in each of the first two words, followed by the four corresponding row indices i. This continues for each column j in turn.

The reader may well ask why go to so much trouble to pack and unpack words if the matrix is being stored out of central memory in virtually limitless backing store. The answer lies with the other bogey of computer users: time. Much of the solution time is spent in transferring information from backing store to central memory, and by reducing the volume of words passing in and out of central memory one is able to achieve significant savings in total running time.

De Buchet (1971) has some interesting figures on operation of the Ophelie code on the CDC 6600 which show an exponential growth in total run time to solve a particular problem with decreasing size of central memory.

The concept of *super-sparsity* has arisen in recent times. This stems from an observation by Kalan (1971) that most LP matrices were "super-sparse" in that not only were the number of nonzero coefficients small, but also the numerical value of many of the coefficients were the same, i.e., the number of *distinct* coefficients is even smaller. For example, many of the coefficients are +1.0, −1.0, +2.0, etc.

Thus we can store the matrix **A** by storing:

1. The row positions for of each nonzero entry column by column, thus accounting for i and j.
2. The address in a table of floating-point numbers that the coefficient has. Thus we store an integer *address* k instead of a floating-point number a_{ij} for each nonzero entry. The same floating-point number may appear several times in the matrix, but is only stored once.

Example As a simple example, consider the following matrix:

$$\mathbf{A} = \begin{bmatrix} 1.5 & 0 & 0 \\ 1.5 & 3.0 & 0 \\ 0 & 0 & 1.5 \end{bmatrix}$$

The array of floating point numbers is

$$\text{TABLE (1)} = 1.5$$
$$\text{TABLE (2)} = 3.0$$

(thus requiring the storage of only two numbers).

Suppose the position information on each nonzero entry is stored as one integer word, using the pattern: $i\,0\,0\,j\,0\,0\,k$

where i = row number

j = column number

k = address in TABLE of coefficients

Then **A** is stored as

1	0	0	1	0	0	1
2	0	0	1	0	0	1
2	0	0	2	0	0	2
3	0	0	3	0	0	1

which requires only four integer words. Again, we may avoid repeating the index j by storing the entries column by column.

If the matrix is at all large and the number of distinct coefficients small, tremendous savings in storage are achieved, even to the extent of being able to store **A** entirely within central memory. The time saved on disk/central memory transfers more than offsets the extra work of packing and unpacking the integer words.

3-3 ROUNDING ERROR

3-3-1 Definition

Floating-point numbers are stored in the computer using essentially scientific notation, i.e., the significant digits and the appropriate power of (say) 10. If the machine stores only the first eight significant digits then the sum of 0.12345×10^{-3} and 1.2345×10^3 (which is 1,234.50012345) is stored as 1.2345001×10^3 and we have lost nearly all the information contributed by the smaller number. This inability to store numbers exactly gives rise to the phenomenon known to numerical analysts as rounding error: the answer to a finite precision arithmetic operation is usually not given exactly. The situation obviously becomes aggravated when a sequence of arithmetic operations is made, since the rounding error "accumulates" in the sense that the possible difference between the exact answer and the stored answer widens.

3-3-2 Avoidance

Our first requirement is to make the elements of the matrix **A** similar in magnitude. A process known as *equilibration* makes the largest element of each row and column the same (say 1.0), and can be accomplished very simply by dividing each row and column in turn by a suitable number. More elaborate methods of scaling the matrix are available, one of the most effective being that due to Hamming (1971). Most commercial LP codes use some means of automatically scaling the matrix before commencing the simplex iterations.

In practice, it turns out that the physical nature of the problem under study usually gives rise to matrix elements which are similar in magnitude, so the importance of scaling is a subject of some dispute. Nevertheless, scaling the matrix is all that can be done a priori, and then the sequence of pivot operations associated with the simplex iterations determine the growth of rounding error as the algorithm proceeds.

3-3-3 Monitoring

To monitor the growth of rounding error, the most commonly used device is to consider the equation

$$\mathbf{Ax} = \mathbf{b} \tag{3-1}$$

The left-hand side is calculated row by row using the *stored* values of $\mathbf{x}(\boldsymbol{\beta}, \mathbf{0})$ and if for some row the difference between that and the corresponding element of the right-hand side $\mathbf{b}$ (called the *row residual*) is not within some error tolerance, then the flag goes up signalling excessive rounding error. We could similarly check the *column residual*

$$\mathbf{c}_B - \mathbf{B}^\top \boldsymbol{\pi}$$

which would likewise be zero in the absence of rounding error.

Recall from Sec. 2-6 that the representation of $\mathbf{B}^{-1}$ is modified at each iteration to reflect the column interchange. This is where the rounding error accumulates, and when the error test fails we must produce a clean $\mathbf{B}^{-1}$. The process is known as *reinversion*, and since it constitutes an important practical part of the solution process we shall consider it at length in Sec. 3-5.

3-3-4 Error Tolerances

The other by-product of the use of floating-point arithmetic and consequent rounding error is the need to use error tolerances instead of testing for zero exactly. Thus we have

- TOLDJ (10^{-5}) The tolerance on reduced cost d_j in Step 3 of the simplex iteration (Sec. 2-6). If d_j is greater than $-$TOLDJ for all nonbasic variables then the solution is optimal.
- TOLAIJ (10^{-10}) If a coefficient a_{ij} is of magnitude less than TOLAIJ it is treated as zero.
- TOLPIV (10^{-8}) Using an element as a pivot (Eq. (2-39)) is avoided if it is of magnitude less than TOLPIV. If all available pivots are less than TOLPIV a reinversion is called for.
- TOLRHS (10^{-6}) The error tolerance discussed in the previous section. If a row of $\mathbf{Ax}$ differs from the corresponding value of $\mathbf{b}$ by more than TOLRHS, excessive rounding error is present.

The above numbers in parentheses are typical values used for a 60-bit floating-point word, giving approximately 15 decimal places accuracy. While they are all tolerances for zero, their different magnitudes reflect the relative importance of the allowable error.

3-4 PRODUCT AND FACTORED FORMS OF THE INVERSE

3-4-1 Product Form

We have already laid the foundation for a sparse representation of $\mathbf{B}^{-1}$ in our previous discussions. Recall:

1. In Sec. 2-6 it was shown that the new $\bar{\mathbf{B}}^{-1}$ after the pivot operation could be represented in the form

$$\bar{\mathbf{B}}^{-1} = \mathbf{E}\mathbf{B}^{-1} \tag{3-2}$$

where

$$\mathbf{E} = \begin{bmatrix} 1 & & & & & \\ & 1 & & & & \\ & & 1 & & & \\ & & & \boldsymbol{\eta}_q & & \\ & & & & 1 & \\ & & & & & 1 \end{bmatrix} = \mathbf{I} + (\boldsymbol{\eta}_q - \mathbf{e}_p)\mathbf{e}_p^{\mathsf{T}} \tag{3-3}$$

2. In Sec. 2-7 it was shown that it is possible to start with an initial basis matrix equal to the identity matrix

$$\mathbf{B} = \mathbf{I} = \mathbf{B}^{-1} \tag{3-4}$$

Thus there is no need to store $\mathbf{B}^{-1}$ as an explicit matrix at all. At the kth iteration $\mathbf{B}^{-1}$ can be represented in the form

$$\mathbf{B}^{-1} = \mathbf{E}_k\mathbf{E}_{k-1} \ldots \mathbf{E}_2\mathbf{E}_1\mathbf{I} \tag{3-5}$$

This is known as the *product form* of the inverse. There are two ramifications of the use of this form

1. The sequence of elementary matrices $\mathbf{E}_1\mathbf{E}_2 \ldots \mathbf{E}_k$ is stored on a so-called *eta file* in which only the vector $\boldsymbol{\eta}_q$ (or $\boldsymbol{\alpha}_q$) and its column position p are recorded for each matrix $\mathbf{E}_k$. Since the vector $\boldsymbol{\eta}_q$ is likely to be sparse if the $\mathbf{A}$ matrix is itself sparse, only the nonzero entries are recorded, in the same manner as the matrix $\mathbf{A}$ (Sec. 3-2), and for large problems is stored out of central memory.
2. The BTRAN and FTRAN operations corresponding to Steps 1 and 4 of the simplex method (Sec. 2-6) are a sequence of simple inner products

BTRAN: Evaluate $\mathbf{c}_B^{\mathsf{T}}\mathbf{B}^{-1}$. Using Eq. (3-5),

$$\mathbf{c}_B^{\mathsf{T}}\mathbf{B}^{-1} = \mathbf{c}_B^{\mathsf{T}}\mathbf{E}_k\mathbf{E}_{k-1} \ldots \mathbf{E}_2\mathbf{E}_1 \tag{3-6}$$

where using Eq. (3-3), each matrix–vector operation is of the form

$$\mathbf{c}_B^{\mathsf{T}}\mathbf{E}_k = \mathbf{c}_B^{\mathsf{T}} + (\mathbf{c}_B^{\mathsf{T}}\boldsymbol{\eta}_q - \mathbf{c}_B^{\mathsf{T}}\mathbf{e}_p)\mathbf{e}_p^{\mathsf{T}} \tag{3-7}$$

i.e., the inner product $\mathbf{c}_B^T\boldsymbol{\eta}_q$ replaces the pth element of $\mathbf{c}_B^T$. The result of Eq. (3-7) is then postmultiplied by $\mathbf{E}_{k-1}$, then $\mathbf{E}_{k-2}, \ldots, \mathbf{E}_1$. Thus, only one element of the vector $\mathbf{c}_B^T$ changes at each step of the sequence.

FTRAN: Evaluate $\mathbf{B}^{-1}\mathbf{a}_q$. Using Eq. (3-5),

$$\mathbf{B}^{-1}\mathbf{a}_q = \mathbf{E}_k\mathbf{E}_{k-1} \ldots \mathbf{E}_2\mathbf{E}_1\mathbf{a}_q$$

Again using Eq. (3-3), we have for each matrix–vector operation

$$\mathbf{E}_1\mathbf{a}_q = \mathbf{a}_q + (\boldsymbol{\eta}_{q'} - \mathbf{e}_p)\mathbf{e}_p^T\mathbf{a}_q \tag{3-8}$$

The result is then premultiplied by $\mathbf{E}_2$, then $\mathbf{E}_3, \ldots, \mathbf{E}_k$. This time the vector $\mathbf{a}_q$ changes at each step. $\mathbf{e}_p^T\mathbf{a}_q$ is just the pth element of the vector $\mathbf{a}_q$, say a_{pq}. This scalar times the vector $\boldsymbol{\eta}_{q'}$ is added to the vector $\mathbf{a}_q$ for all but the pth position, where a_{pq} times the pth element of $\boldsymbol{\eta}_{q'}$ replaced a_{pq}. (Note that both $\mathbf{a}_q$ and $\boldsymbol{\eta}_{q'}$ will usually be sparse.)

We must quit the process after a while as the number of simplex iterations k becomes too large. The sequence of inner-product calculations gives rise to rounding error, and the error checks discussed in the previous section are one criterion for stopping. Another reason is the simple fact that if k is too large we waste too much time passing the eta file back and forward between backing store and central memory. Thus another stopping criterion is when k exceeds, say, $m + 50$, although commercial codes use the internal clock of the computer, and when the marginal iteration rate (iterations per unit time) starts to fall away due to increased disk/central memory transfer time, a reinversion is called for.

Having stopped the sequence of elementary transformations, we must now clean up our representation of $\mathbf{B}^{-1}$ (*reinvert*) and it is worth emphasizing here that this can be accomplished with (at most) m elementary transformations, not k, starting from our original initial basis $\mathbf{B} = \mathbf{I}$.

3-4-2 LU Factorization of Basis

We have noted in the previous section that the two foes, storage and rounding error, reappear to limit the extent to which the product form of the inverse can be continued.

Factorization methods involve transforming the basis into the product of two triangular matrices

$$[\mathbf{B}] = \mathbf{L}\;\mathbf{U} \tag{3-9}$$

where **L** is lower triangular and **U** is upper triangular.

Thus solution of the FTRAN equation (cf. Eq. (2-33))

$$\mathbf{B}\boldsymbol{\alpha}_q = \mathbf{a}_q \tag{3-10}$$

can be accomplished by solving

$$\mathbf{L}\mathbf{y} = \mathbf{a}_q \qquad \text{for } \mathbf{y} \tag{3-11}$$

by forward-substitution then solving

$$\mathbf{U}\boldsymbol{\alpha}_q = \mathbf{y} \qquad \text{for } \boldsymbol{\alpha}_q \tag{3-12}$$

by back-substitution.

Historically, the *elimination form of the inverse* (EFI), due to Markowitz (1957) was the first **LU** factorization method and was introduced to preserve sparsity during reinversion, i.e., refactorizing $\mathbf{B} = \mathbf{LU}$. However, once reinversion was completed further pivot operations were handled using the ordinary product form (PFI).

Bartels and Golub (1969) proposed *updating* **L** and **U** in a numerically stable way. To consider what happens when a column interchange occurs, it is convenient to express Eq. (3-9) in the form

$$\mathbf{L}^{-1}\mathbf{B} = \mathbf{U} \tag{3-13}$$

Thus when $\mathbf{a}_q$ replaces the pth column of **B** to form $\bar{\mathbf{B}}$ (Eq. (2-35)), we get

$$\mathbf{L}^{-1}\bar{\mathbf{B}} = \tag{3-14}$$

where

$$\gamma = \mathbf{L}^{-1}\mathbf{a}_q \tag{3-15}$$

has correspondingly replaced the pth column of **U**.

On permuting the columns so that γ is last, and moving the other columns up one place, we get

$$\mathbf{L}^{-1}\bar{\mathbf{B}} =$$

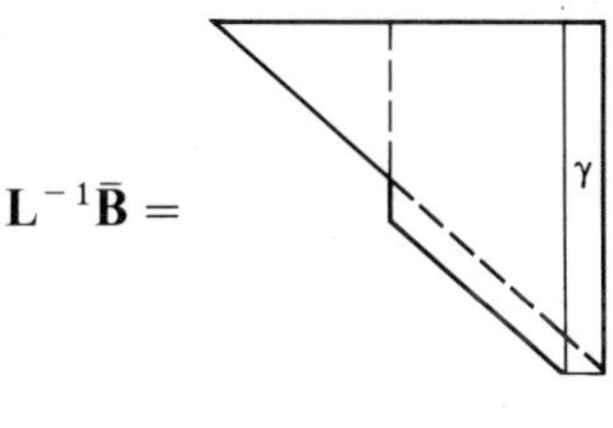

$$= \mathbf{H} \text{ say.}$$

H is called an upper-Hessenberg matrix. The permutation matrix is deliberately left out: it suffices to keep a record of the column positions.

Now the Bartels–Golub method consists of a series of elementary transformations which reduce the subdiagonal elements of **H** to zero. The subdiagonal elements are

$$\begin{array}{lll} u_{p+1\,p+1} & \text{in column} & p \\ u_{jj} & \text{,, \quad ,,} & j-1 \\ u_{mm} & \text{,, \quad ,,} & m-1 \end{array}$$

where u_{ij} is the ijth element of the *original* **U**,

i.e., we have $\mathbf{L}^{-1}\bar{\mathbf{B}} =$ (3-16)

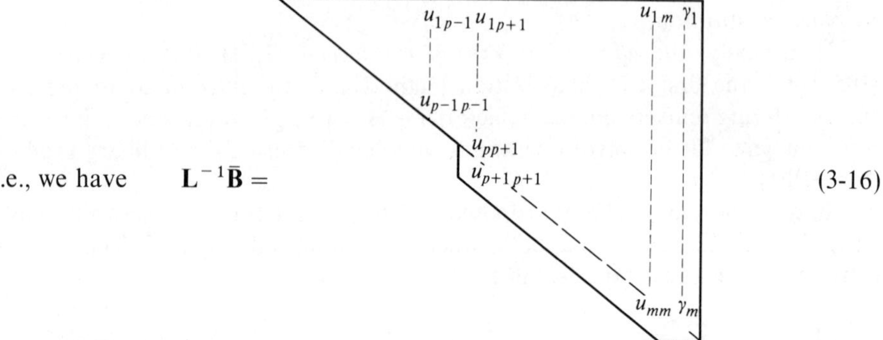

We wish to reduce the subdiagonal elements $u_{p+1\,p+1}, \ldots, u_{mm}$ to zero.

The Bartels–Golub method accomplishes this by a series of elementary transformations

$$\mathbf{M}_j, j = p+1, \ldots, m$$

so that

$$\mathbf{M}_m \ldots \mathbf{M}_j \ldots \mathbf{M}_{p+1}[\mathbf{L}^{-1}\bar{\mathbf{B}}] = \bar{\mathbf{U}} \qquad \text{where } \bar{\mathbf{U}} \text{ is upper triangular.} \quad (3\text{-}17)$$

Each of the transformations works on *only two elements* (the diagonal element and the subdiagonal element) of each column in turn.

Therefore we construct transformations

$$\mathbf{M}_{p+1}, \mathbf{M}_{p+2}, \ldots, \mathbf{M}_m$$

of the form

$$\mathbf{M}_j = \begin{bmatrix} 1 & & & & & & & & \\ & 1 & & & & & & & \\ & & 1 & & & & & & \\ & & & 1 & & & & & \\ & & & & [\tilde{\mathbf{M}}_j] & & & & \\ & & & & & 1 & & & \\ & & & & & & 1 & & \\ & & & & & & & 1 & \\ & & & & & & & & 1 \end{bmatrix} \qquad (3\text{-}18)$$

where the 2 × 2 matrices

$$\tilde{\mathbf{M}}_j, j = p + 1, p + 2, \ldots, m$$

occupy columns $j - 1$ and j and are of the form

$$\tilde{\mathbf{M}}_j = \begin{bmatrix} 1 & 0 \\ m_j & 1 \end{bmatrix} \quad \text{if no row interchanges, } m_j = -u_{jj}/\bar{u}_{j-1j} \tag{3-19}$$

or

$$\tilde{\mathbf{M}}_j = \begin{bmatrix} 0 & 1 \\ 1 & m_j \end{bmatrix} \quad \text{if rows interchange, } m_j = -\bar{u}_{j-1j}/u_{jj} \tag{3-20}$$

($\bar{u}_{j-1,j}$ is the original $u_{j-1,j}$ premultiplied by $\mathbf{M}_{p+1}, \ldots, \mathbf{M}_{j-1}$.) We thus have the following results:

No row interchanges

$$\mathbf{M}_j\mathbf{u}_j = \begin{bmatrix} 1 & & & & & & & \\ & 1 & & & & & & \\ & & 1 & & & & & \\ & & & 1 & 0 & & & \\ & & & -\dfrac{u_{jj}}{\bar{u}_{j-1j}} & 1 & & & \\ & & & & & 1 & & \\ & & & & & & 1 & \\ & & & & & & & 1 \end{bmatrix} \begin{bmatrix} \bar{u}_{1j} \\ \vdots \\ \bar{u}_{j-1j} \\ u_{jj} \\ 0 \\ \vdots \\ 0 \end{bmatrix}$$

$$= \begin{bmatrix} \bar{u}_{1j} \\ \vdots \\ \bar{u}_{j-1j} \\ \leftarrow \\ 0 \\ \vdots \\ 0 \end{bmatrix} -\frac{u_{jj}}{\bar{u}_{j-1j}} \cdot \bar{u}_{j-1j} + u_{jj} = 0 \tag{3-21}$$

(Note that $\mathbf{u}_j$ occupies column $j - 1$ of $\mathbf{H}$.)

Example $j = p + 1$

$\mathbf{H} =$ 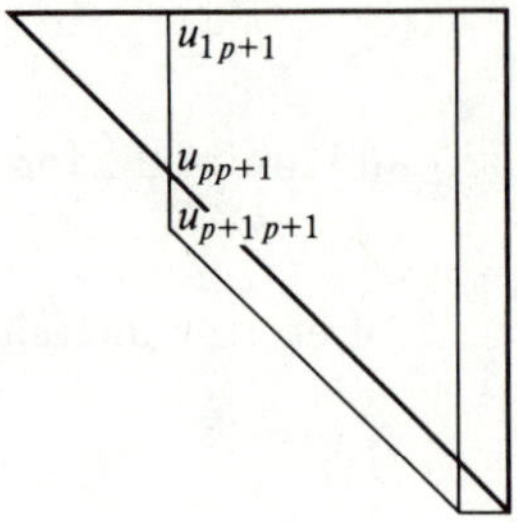

(3-22)

$$\mathbf{M}_{p+1}\mathbf{u}_{p+1} = \begin{bmatrix} 1 & & & & & & \\ & 1 & & & & & \\ & & \begin{bmatrix} 1 & 0 \\ -\dfrac{u_{p+1\,p+1}}{u_{pp+1}} & 1 \end{bmatrix} & & & & \\ & & & 1 & & & \\ & & & & 1 & & \\ & & & & & & 1 \end{bmatrix} \begin{bmatrix} u_{1\,p+1} \\ \vdots \\ u_{pp+1} \\ u_{p+1\,p+1} \\ 0 \\ \vdots \\ 0 \end{bmatrix}$$

$$= \begin{bmatrix} u_{1\,p+1} \\ \vdots \\ u_{pp+1} \\ 0 \\ \vdots \\ 0 \end{bmatrix} \begin{matrix} \left.\vphantom{\begin{matrix} u \\ \vdots \\ u \end{matrix}}\right\}\text{unchanged} \\ \leftarrow -\dfrac{u_{p+1\,p+1}}{u_{pp+1}}u_{pp+1} + u_{p+1\,p+1} = 0 \\ \\ \end{matrix} = \begin{bmatrix} u_{1\,p+1} \\ \vdots \\ u_{pp+1} \\ 0 \\ 0 \\ \vdots \\ 0 \end{bmatrix} \quad (3\text{-}23)$$

With row interchanges

$$\mathbf{M}_j\mathbf{u}_j = \begin{bmatrix} 1 & & & & & & \\ & 1 & & & & & \\ & & 1 & & & & \\ & & & \begin{bmatrix} 0 & 1 \\ 1 & -\dfrac{\bar{u}_{j-1\,j}}{u_{jj}} \end{bmatrix} & & & \\ & & & & 1 & & \\ & & & & & 1 & \\ & & & & & & 1 \end{bmatrix} \begin{bmatrix} u_{1j} \\ \vdots \\ \bar{u}_{j-2\,j} \\ \bar{u}_{j-1\,j} \\ u_{jj} \\ 0 \\ \vdots \\ 0 \end{bmatrix}$$

$$= \begin{bmatrix} \bar{u}_{1j} \\ \vdots \\ \bar{u}_{j-2j} \\ u_{jj} \\ \quad \longleftarrow \bar{u}_{j-1j} - \dfrac{\bar{u}_{j-1j}}{u_{jj}} \cdot u_{jj} = 0 \\ 0 \\ \vdots \\ 0 \end{bmatrix} \tag{3-24}$$

Similarly, $\mathbf{M}_j$ applied to columns $j + 1, \ldots, m$ will also effect a row interchange between $j - 1$ and j.

Bartels and Golub maintain $|m_j| \leqslant 1$ (i.e., if $|u_{jj}| > |\bar{u}_{j-1j}|$ then interchange) for reasons of numerical stability (i.e., they put the largest element of the two into the diagonal position). Comparison of Eqs. (3-17)–(3-24) with Eqs. (1-54)–(1-62) indicates that this strategy is equivalent to gaussian elimination with partial pivoting. We could relax this requirement to the form

$$|m_j| \leqslant 1/\mu$$

where $0 \leqslant \mu \leqslant 1$ is specified a priori in order to give greater freedom in reducing the number of new nonzeros created. Thus we have after applying the sequence,

$$\mathbf{M}_m \ldots \mathbf{M}_{p+1} \cdot \mathbf{L}^{-1}\bar{\mathbf{B}} = \bar{\mathbf{U}}$$

$$= \tag{3-25}$$

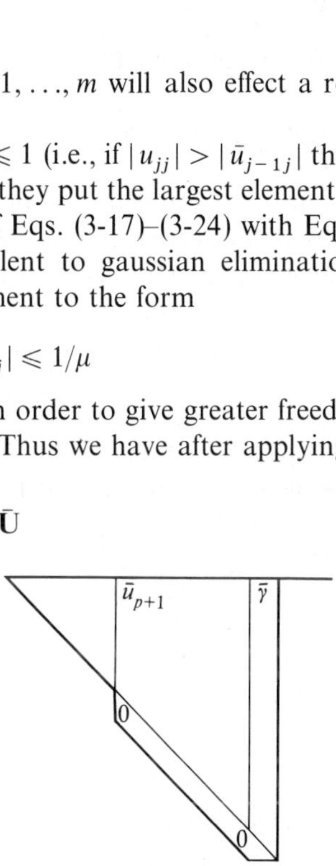

which is upper triangular.

Note that, as in the product form of the inverse, there is no need to compute $\mathbf{L}^{-1}$ itself explicitly, but we could start with $\mathbf{B} = \mathbf{B}^{-1} = \mathbf{U} = \mathbf{I}$, corresponding to an all-slack basis. In fact, it is not necessary to start with $\mathbf{B} = \mathbf{I}$; it would suffice to start with $\mathbf{B} = \mathbf{L}$, $\mathbf{U} = \mathbf{I}$. We can then solve the equation as required ($\mathbf{L}^{-1}$ is never computed). Getting an initial factorization $\mathbf{B} = \mathbf{LU}$ is discussed further in Sec. 3-5. $\mathbf{U}$ is stored explicitly; the implied permutations can be recorded merely by keeping a list of the column positions.

Since the terms $\mathbf{M}_j$ are stored as a sequence of elementary transformations, information required to generate these (the vector and the column position p) can be held in a sequential disk file as an eta file. However, since row interchanges may or may not occur in $\mathbf{U}$, nonzeros can be created, so $\mathbf{U}$ should be held wholly within central memory as it becomes rather complicated to allow for new nonzeros being inserted in a sequential file being held out of central memory on disk or tape.

3-4-3 The Forrest and Tomlin Method

The *advantage* of the Bartels–Golub Method is that it is numerically stable. The *disadvantages* are:

1. New nonzeros may be created in $\mathbf{U}$. We therefore may lose sparseness by fill-in of nonzeros.
2. We need to store $\mathbf{U}$ within central memory.

The Forrest and Tomlin method (Forrest and Tomlin, 1972) was designed to overcome these two disadvantages, at some sacrifice in numerical stability. In this procedure, no new nonzeros are created in $\mathbf{U}$, therefore $\mathbf{U}$ can be stored out of central memory, in a sequential disk-file.

Let $\mathbf{r}$ be the vector defined by the equation

$$\mathbf{U}^{\mathsf{T}}\mathbf{r} = \sigma\mathbf{e}_p \tag{3-26}$$

where σ is some appropriate scalar (say u_{pp}).

Define $\mathbf{R}$ as the elementary transformation

$$\mathbf{R} = \mathbf{I} + \mathbf{e}_p(\mathbf{r}^{\mathsf{T}} - \mathbf{e}_p^{\mathsf{T}}) \tag{3-27}$$

Then
$$\mathbf{RU} = \mathbf{U} + \mathbf{e}_p(\mathbf{r}^{\mathsf{T}}\mathbf{U} - \mathbf{e}_p^{\mathsf{T}}\mathbf{U}) \tag{3-28}$$

but
$$\mathbf{r}^{\mathsf{T}}\mathbf{U} = \sigma\mathbf{e}_p^{\mathsf{T}} \qquad \text{by definition} \tag{3-29}$$

and $\mathbf{e}_p^{\mathsf{T}}\mathbf{U}$ is the pth row of $\mathbf{U}$.

We have $\mathbf{RU} =$ (3-30)

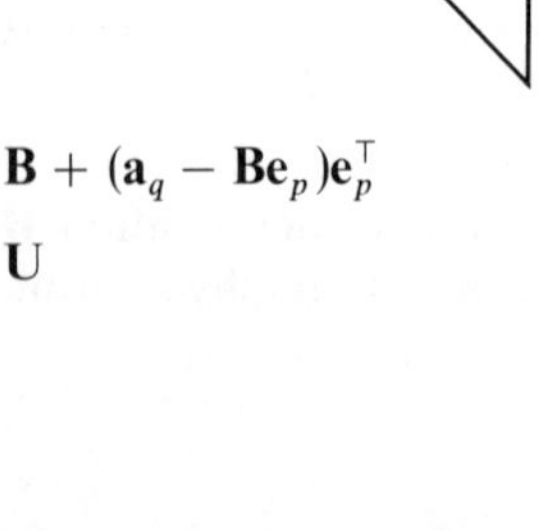

u_{1p}

$u_{p-1\,p}$

0 --- 0

u_{pp}

if $\sigma = u_{pp}$

If
$$\bar{\mathbf{B}} = \mathbf{B} + (\mathbf{a}_q - \mathbf{B}\mathbf{e}_p)\mathbf{e}_p^{\mathsf{T}} \tag{3-31}$$

and
$$\mathbf{L}^{-1}\mathbf{B} = \mathbf{U} \tag{3-32}$$

(as in the previous section),

then $\mathbf{RL}^{-1}\bar{\mathbf{B}} =$

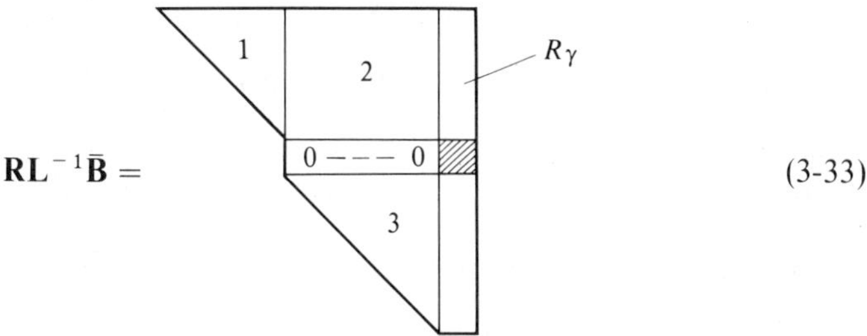

(3-33)

with the same permutation as in the Bartels–Golub method.

Thus, with one more permutation, which puts the row of zeros on the bottom and shifts the others up, we have

$\mathbf{RL}^{-1}\bar{\mathbf{B}} =$ (3-34)

which is upper triangular.

Note that all we have done to the rest of $\mathbf{U}$ is delete the pth row. We have added *no* new zeros to the rest of $\mathbf{U}$. Again, we have ignored the permutation matrices. We can now store $\mathbf{U}$ out of central memory as a sequence of column vectors with an accompanying register of deletions and permutations. The elementary transformations $\mathbf{R}$ can of course be held as an "eta"-type file.

The Forrest and Tomlin method is theoretically equivalent to the Bartels–Golub method with a row interchange at each step $j = p + 1, \ldots, m$. If a row interchange occurs at each step, the cumulative effect is to bring each row up one place and no new nonzeros are created. However, the lack of choice on interchanging makes it less numerically stable than the Bartels–Golub method, and this is the price paid for the savings in storage.

Recently Saunders (1976) has shown how most of $\mathbf{U}$ can be held out of central memory in the Bartels–Golub method, so this approach solves both the rounding error and the storage problem. We shall defer discussion of this until Sec. 3-5-1.

3-4-4 Other Factorizations

There are a number of possible factorizations of the basis matrix. Gill and Murray (1973) proposed the factorization

$$\mathbf{B} = \mathbf{LQ} \tag{3-35}$$

where **L** is lower triangular and **Q** is *orthogonal*, i.e., it possesses the properties

$$\mathbf{Q}^{\top}\mathbf{Q} = \mathbf{Q}\mathbf{Q}^{\top} = \mathbf{I} \tag{3-36}$$

This factorization possesses many desirable numerical properties in terms of avoiding rounding error, but does create rather more nonzeros than the **LU** factorization.

We do not need to store **Q**, for we can solve the BTRAN equation

$$\mathbf{B}^{\top}\boldsymbol{\pi} = \mathbf{c}_B \tag{3-37}$$

by calculating

$$\mathbf{y} = \mathbf{B}\mathbf{c}_B \tag{3-38}$$

Thus we solve

$$\mathbf{B}\mathbf{B}^{\top}\boldsymbol{\pi} = \mathbf{L}\mathbf{Q}\mathbf{Q}^{\top}\mathbf{L}^{\top}\boldsymbol{\pi} = \mathbf{L}\mathbf{L}^{\top}\boldsymbol{\pi} = \mathbf{y} \tag{3-39}$$

for $\boldsymbol{\pi}$.

For the FTRAN equation, $\mathbf{B}\boldsymbol{\alpha}_q = \mathbf{a}_q$, we have

$$\mathbf{L}\mathbf{u} = \mathbf{a}_q, \qquad \mathbf{L}^{\top}\mathbf{v} = \mathbf{u}, \qquad \boldsymbol{\alpha}_q = \mathbf{B}^{\top}\mathbf{v} \tag{3-40}$$

Saunders (1972) considered treating **L** as a sparse matrix and gave a version in which **L** is stored as a sequence of elementary transformations.

The price we pay for not having to store **Q** is the extra work in computing **y** from Eq. (3-38) (and similarly in Eq. (3-40)). In a later paper, Gill, Murray, and Saunders (1975) have shown how both **L** and **Q** can be updated in product form.

This factorization is of importance in solving large structured systems since it turns out that the sparsity in $\mathbf{B}\mathbf{B}^{\top}$ induced by the structure is not affected by the rearranging of the columns of **B** dictated by the simplex steps.

3-5 REINVERSION

The reinversion procedure is probably the most well-polished part of all commerical LP codes, since the new representation of the inverse must be as error-free and sparse as possible to keep the subsequent iterations fast and accurate.

At the time of reinversion, which has been signalled by excess rounding error or growth in nonzeros (too large an eta file), we know which variables are in the basis. However, we do *not* take the corresponding m columns of **A** which form **B** and calculate its inverse explicitly. Rather, we could start with our initial basis equal to **I**, corresponding to slack or artificial variables, and perform m pivot operations in which we sequentially insert the correct columns of **B**. Note that this way it would only take m operations to restore the inverse (or factorization) to where we were after k iterations. This means that $(k - m)$ variables have entered *and left* the basis and thus could be regarded as wasted effort; however, we were not to know that a priori. (As we shall see, if we adopt the factorization approach discussed in Secs. 3-4-2 and 3-4-3, we will usually require much less than m operations to factorize $\mathbf{B} = \mathbf{LU}$).

The high state of development of reinversion procedures arises as a consequence of the fact that we can put the columns (corresponding to the current basic variables) into **B** in any order, i.e., we can rearrange the columns of **B** in any order. We can also rearrange the rows of **B** in any order, since this simply corresponds to a shuffling of the constraints. We only need to keep a record of current row and column positions.

Deciding the row and column order of the basis **B** means essentially deciding at each step which column q to bring into the basis (cf. Step 3, Sec. 2-6) and which row p to pivot on (cf. Step 5, Sec. 2-6), except we do not use the simplex rules. Now, since there is considerable freedom of choice we must have some decision rules for selection.

For minimum rounding error, the criterion would be to choose p and q so that the pivot element (α_{pq} in Eq. (2-39)) is as large as possible. However, this would be much too expensive when **B** is large.

For sparsity, we wish to create as few nonzero entries in our representation as possible. To some extent this takes priority over rounding error; we would only reject a pivot chosen for sparsity if the resulting pivot element α_{pq} was too small (below, say, $10^{-3}\ \alpha_{pq}^{\max}$ in magnitude, where $\alpha_{pq}^{\max}$ is the maximum pivot in the column).

It has been reported (Dantzig *et al.*, 1969) that without any attempt to sort the rows and columns, the product-form inverse yields a 5–10 fold increase in nonzero elements in problems in which **B** itself is 2–5 percent dense, and this is cut dramatically by even a partial sort. Such figures illustrate the immense practical importance of this part of the procedure.†

One of the most successful decision rules in earlier codes was that due to Markowitz (1957): select that element as pivot, such that the product of the number of other nonzero elements in its row, times the number of other nonzeros in its column, is a minimum. This product is the number of multiplications/additions and also indicates the maximum fill-in which may occur from that pivot operation.

This has been superseded more recently by the procedure of Hellerman and Rarick (1971, 1972) which will be described in general terms in the next section.

3-5-1 The Partitioned Preassigned Pivot Procedure (P^4)

P^4 is essentially a method of making **B** as near to lower triangular as possible. Clearly, if **B** *was* lower triangular then *no* new nonzeros would be created in either the product form or the **LU** factorization (indeed $\mathbf{B} = \mathbf{L}$, $\mathbf{U} = \mathbf{I}$). Such an ideal

† Even without sorting, the **LU** factorization of **B** is much more sparse than the product form (PFI), which gives precisely the numbers in **L** and $\mathbf{U}^{-1}$ (not **U**). This fact gave rise to Markowitz suggesting the elimination form (EFI) for reinversion.

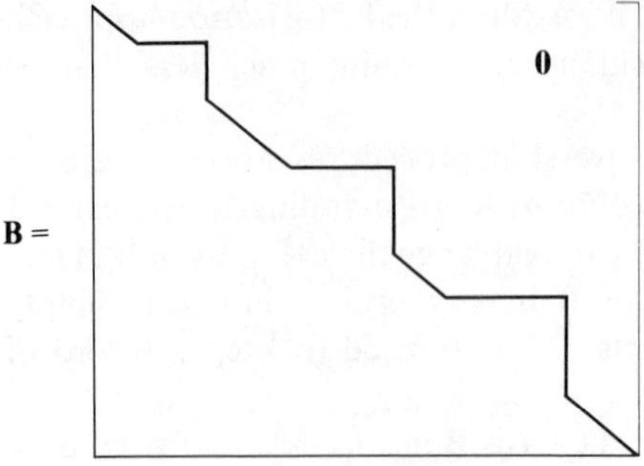

Figure 3-1 Block lower-triangular system.

state rarely exists and the best we can hope for is a *block-triangular* system, which is lower triangular except for certain blocks along the diagonal (see Fig. 3-1). The steps of P^4 are as follows:

1. Remove the row and column singletons (rows and columns with just 1 element).
2. Permute the remaining portion of the matrix to block triangular form.
3. Process each block ("bump") to minimize fill-in within each block.

Step 1 requires permuting the rows and columns so that the singletons are along the diagonal. The row singletons are sorted toward the top left and column singletons are sorted toward the bottom right. Each time a singleton is identified and sorted, the corresponding row and column are removed from consideration; we thus look for a singleton in the *remaining* matrix. Thus **B** will take the form as shown in Fig. 3-2 once Step 1 is completed.

Step 2 is a little more involved. The remaining submatrix $\mathbf{B}_1$ can be further

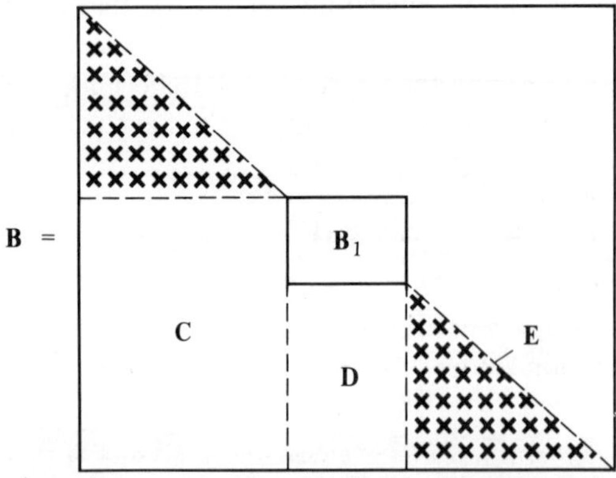

Figure 3-2 System after removal of singletons.

reduced to a series of smaller blocks ("bumps"). The columns in part **C** of **B** cannot be rearranged (although the *rows* corresponding to the rows in $\mathbf{B}_1$ can be). Similarly, the rows in part **E** of **B** cannot be rearranged (although the *columns* corresponding to the columns in $\mathbf{B}_1$ can be). Thus we are able to permute both the rows and columns of $\mathbf{B}_1$.

The reduction of $\mathbf{B}_1$ to block-triangular form takes place in two phases:

1. The columns are ordered so that there are nonzero entries along the diagonal. This process is known as finding a transversal. This gives a tentative pivot assignment.
2. The rows and columns are permuted symmetrically. This means that, if we visualize each equation being solved for the diagonal variable, the variables are solved with the same equations as in the tentative assignment, but the *order* in which they are solved is altered. By rearranging the order of solution, it is usually possible to find equations with no superdiagonal terms, i.e., the variable being solved is expressed entirely in terms of variables which have previously been calculated. This partitions one block into two smaller blocks, separated by a diagonal element.

Thus we eventually end up with $\mathbf{B}_1$ taking the form as shown in Fig. 3-3. It should be mentioned that while the task of permuting to block-triangular form is conceptually straightforward, a vast amount of research (quite apart from that of Hellerman and Rarick) has been undertaken towards reducing the computational complexity of accomplishing it; and a number of algorithms based on graph theory have been proposed (see for example Duff, 1976). Duff has shown that the resulting block-triangular form is unique with regard to the number of blocks and the rows and columns in each column, except for their ordering within the block.

Step 3 examines each block in turn. Hellerman and Rarick proposed a

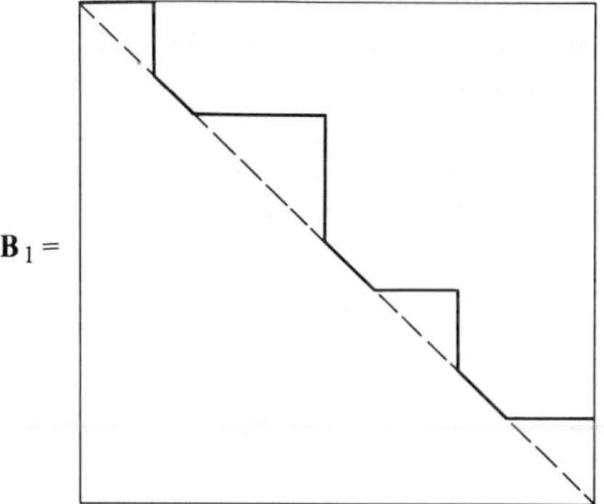

Figure 3-3 Partitioning of $\mathbf{B}_1$ into smaller bumps.

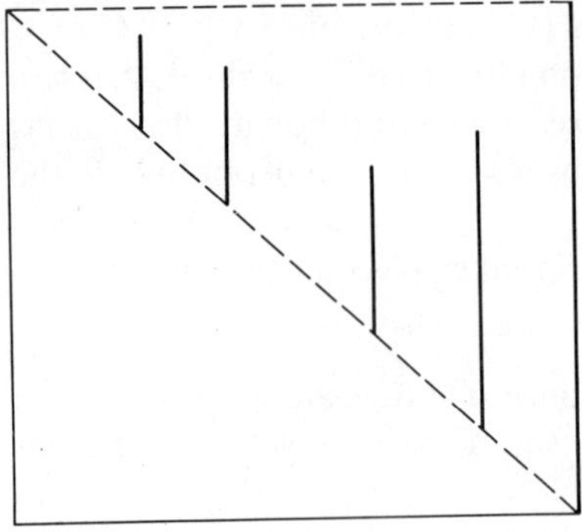

Figure 3-4 Spikes within a bump.

heuristic procedure for sorting each individual block ("bump") into the form of Fig. 3-4, with a number of columns with nonzeros above the diagonal ("spikes").

Note that the largest spike is always the right-hand edge of the block, since this defines the height of the block. Note also that the **LU** factorization will only produce new nonzero entries (i.e., fill-in) in the spike columns and then only as high as each spike. Thus, the ordering is based on the criteria of not only the number of spikes, but also of their height. The basis of the approach of Hellerman and Rarick is the use of a *tally function*, $t_k(j)$, which is defined as the number of nonzeros that column j has in rows whose row counts (nonzeros) are k or less. Thus we examine the available columns (initially all those in the block) and set k equal to the minimum row count (initially 2 or more). The column, j, with the highest tally $t_k(j)$, >1, is selected as a spike and removed. If the highest $t_k(j) = 1$, this is not adequate to select a spike (since we would not reduce the row count for any other row with k or less nonzeros by removing the column), so the subset of columns which had $t_k(j) = 1$ is re-examined with k increased by 1.

Once a spike is selected, it is placed in a last-in-first-out list, and the remaining columns are examined for the possibility of either:

1. Continuation of the forward triangle (minimum row count of 1); or
2. Further spike selection (minimum row count of 2 or more).

For case 1, if $t_1(j)$ is 2 (or more) for a particular column j then we are able to place a spike (or more) after j using the last-in-first-out rule.

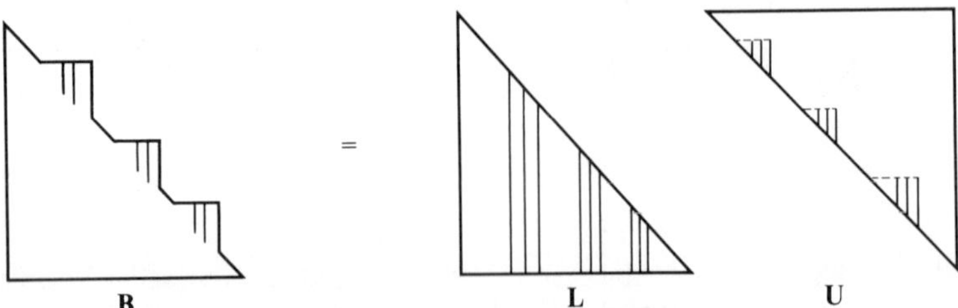

Figure 3-5 LU factorization of **B**.

Ties in the above operations are resolved in favor of removing the most nonzeros from the remaining matrix. The reader may wish to verify that this procedure transforms the matrix below to one with two spikes (Hellerman and Rarick, 1972).

	1	2	3	4	5	6
1	×					×
2	×	×		×	×	×
3	×		×			
4	×		×	×	×	
5				×	×	
6		×	×	×		×

	6	3	5	4	2	1
1	×					×
3		×				×
4		×	×	×		×
5			×	×		
6	×	×		×	×	
2	×		×	×	×	×

Note that we need to record the column order and row order (pivot sequence) as well as an indication of the height of each spike and the column position where it occurs.

Thus an arbitrary basis matrix can be made to look like Fig. 3-5 with its corresponding **LU** factorization. The vertical lines in **L** and **U** in Fig. 3-5 indicate the columns where fill-in may occur, corresponding to the spikes of **B**. The columns of **L** corresponding to the non-spike columns of **B** can be stored by merely a register of the appropriate columns of **A**. The spike columns give rise to elementary transformations of the type discussed in Secs. 3-4-2 and 3-4-3.

The number of spikes in a basis of 1,000 rows is typically between 0 and 100 (although some problems have many more). To maintain numerical stability during $\mathbf{B} = \mathbf{LU}$, any given column inside a bump may have to be swapped with one of the remaining spikes in the same bump. This usually increases the final number of spikes slightly.

It turns out in practice that the effort involved in shuffling the basis matrix around in this manner pays handsome dividends in cutting down on the growth of nonzeros during reinversion, which of course affects both running time and storage for the subsequent simplex iterations.

Saunders (1976) has exploited the form of Fig. 3-5 in updating the **LU** factorization with the Bartels–Golub method.

Consider the upper-triangular matrix **U** of Fig. 3-5. Saunders proposed a column permutation **P** which moved all the spikes to the right-hand side of **U** without changing their order, followed by the same permutation on the rows, to yield

$$\tilde{\mathbf{U}} = \mathbf{P}^{\mathrm{T}}\mathbf{U}\mathbf{P} = \left[\begin{array}{c|c} \mathbf{I} & \mathbf{R} \\ \hline \mathbf{0} & \mathbf{F} \end{array}\right]$$

The form of $\tilde{\mathbf{U}}$ corresponding to Fig. 3-5 would be as shown in Fig. 3-6.

Now when the pth column of **B** is replaced by $\mathbf{a}_q$, Saunders' procedure entails the following steps to update **L** and $\tilde{\mathbf{U}}$:

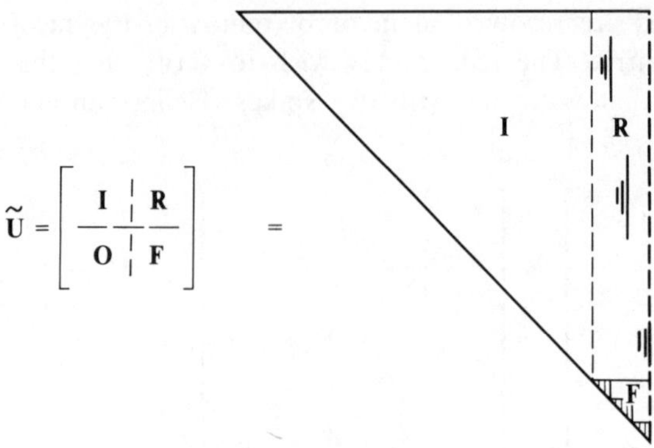

Figure 3-6 Rearrangement of spikes in $\tilde{\mathbf{U}}$.

1. Delete the pth column of $\tilde{\mathbf{U}}$, say $\mathbf{u}_p$, and move the other columns, $p+1, \ldots, m$, up by one.
2. Add $\gamma = \mathbf{L}^{-1}\mathbf{a}_q$ to the end of $\tilde{\mathbf{U}}$.
3. Extract the pth row of $\tilde{\mathbf{U}}$, say $\mathbf{r}^{\mathrm{T}}$, and permute it to the bottom, moving the other rows $p+1, \ldots, m$ up by one (thus preserving the upper-triangular form in columns $1, \ldots, m-1$).

   Now $\mathbf{r}^{\mathrm{T}}$ has nonzero elements only in the column positions corresponding to $\mathbf{R}$ (or $\mathbf{F}$ if $\mathbf{u}_p$ was a spike) as we have already removed the pth column (and thus the diagonal 1.0) in Step 1.
4. Reduce $\mathbf{r}^{\mathrm{T}}$ to a multiple of the unit vector $\mathbf{e}_m^{\mathrm{T}}$ using gaussian elimination with row interchanges, as in the Bartels–Golub method.

Thus, new nonzeros that may occur during Step 4 can only be created in **F**, so **R** may be held out of central memory. **F** is usually sufficiently small to be held within central memory as a dense matrix, so creation of new nonzeros presents no difficulty. For problems with large numbers of spikes, **F** could be stored as a sparse matrix using ordered lists of nonzero entries.

3-6 PRICING METHODS

3-6-1 Introduction

This section is concerned with Steps 2 and 3 of the simplex algorithm (Sec. 2-6). Recall that in Step 2 we evaluate

$$z_j = \boldsymbol{\pi}^{\mathrm{T}}\mathbf{a}_j \tag{3-41}$$

then the reduced cost

$$d_j = z_j - c_j \tag{3-42}$$

for each nonbasic column of the matrix **A**.

Step 3 involves choosing an appropriate nonbasic column with negative reduced cost to enter the basis. Since we are at liberty to choose any such nonbasic column, we do not necessarily need to choose the one with the most negative reduced cost. This gives rise to other methods of selection, which are discussed in Secs. 3-6-4 and 3-6-5. Sections 3-6-2 and 3-6-3 discuss ways of avoiding work when there are a large number of nonbasic columns, stored out of central memory on some form of auxiliary storage, such as disk. The fact that disk/central memory transfers take a significant amount of time leads us to consider implementing the steps of the simplex method in a different way.

3-6-2 Partial Pricing

The most simple device for cutting down on disk/central memory transfer is to avoid computing the reduced cost of *all* the nonbasic variables. Thus we scan only a portion of the nonbasic columns selecting that with (say) the most negative reduced cost. At the next iteration we scan another portion, continuing from where we stopped previously. Thus, not all of the matrix needs to be passed through central memory at each iteration. This achieves a direct saving in disk/central memory transfer time, which is somewhat offset by the fact that since column selection is based on only a portion of the matrix, the total number of iterations will usually increase.

3-6-3 Multiple Pricing

Another way of reducing disk/core transfer time is to select a number (say 5) of columns with significantly negative reduced costs, and process them all in central memory.

Thus we calculate $\boldsymbol{\alpha}_q$ (Step 4 of the simplex method, Sec. 2-6) with all five columns. Of course only one of them enters the basis, but for the next iteration we only consider the remaining four columns as candidates. Since the information is held within central memory for these four, we avoid any disk/central memory transfers. This process is repeated until the remaining columns do not have a sufficiently large negative reduced cost to enter the basis.

The fact that we have $\boldsymbol{\alpha}_q$ for all five columns gives us an extra degree of flexibility. We can choose the entering column q so that the resultant pivot row p (Step 5) yields the highest increase in the objective function. (Since the increase is measured by $-d_q(\beta_p/\alpha_{pq})$, the most negative d_j is not necessarily the best.) This process is known as *suboptimization*. To summarize:

1. Modify Step 3 to choose, say, five candidate columns.
2. Complete Steps 4 and 5 for each of the five candidates.
3. Choose as the entering column the candidate with the largest negative value of $d_q(\beta_p/\alpha_{pq})$.

Once the five $\boldsymbol{\alpha}$'s are computed and saved, only one $\mathbf{E}_k$ needs to be applied to the next iteration to update them, so access to both $\mathbf{A}$ *and* $\mathbf{B}^{-1}$ is avoided.

3-6-4 The Devex Pricing Method

The criterion of reduced cost is equivalent to choosing the largest gradient in the space of the current nonbasic variables, which is a reference space which changes at each iteration. The *Devex* method, developed by Harris (1973), seeks to choose the incoming variable on the basis of largest gradient in the space of the initial nonbasic variables. These will be the structural variables if we start with an all-slack or artificial-variable basis, and of course the structural variables are the ones with the most physical significance. The gradients in the constant reference space are approximated by weighting factors applied to each of the reduced costs.

Starting with weighting factors $T_j = 1$ for all columns, and the initial nonbasic variables as the reference space, suppose that after a number of iterations some of the initial basic variables (nonreference) are now nonbasic in the set of columns C. Also, some of the initial nonbasic variables (reference) are now basic and correspond to the set of rows R (in Eq. (2-17)).

Now when we consider increasing some nonbasic variable x_j with reduced cost d_j, its gradient in the reference space is d_j/T_j, where

$$T_j = \sqrt{\delta_j + \sum_{i \in R} \alpha_{ij}^2} \tag{3-43}$$

$$\boldsymbol{\alpha}_j = \mathbf{B}^{-1}\mathbf{a}_j \tag{3-44}$$

and

$$\delta_j = \begin{cases} 1 & j \notin C \\ 0 & j \in C \end{cases} \tag{3-45}$$

The reference variables basic on rows $i \in R$ will increase by α_{ij} per unit increase in x_j and δ_j has unity only if x_j is in the reference space; thus per unit increase in x_j, the increase in objective is d_j and the distance moved in the reference space is given by Eq. (3-43).

Now since we do not calculate $\boldsymbol{\alpha}_j$ (Step 4) until j is chosen (Step 3), the exact value of T_j is approximated by recursion from the previous iteration (using p and q from the previous iteration)

$$T_j(\text{new}) = \begin{cases} \max\{T_j, |\alpha_{pj}| T_q\}, q \notin C \\ \max\{T_j, |\alpha_{pj}| \sqrt{1 + T_q^2}\}, q \in C \end{cases} \tag{3-46}$$

Thus we only need $\alpha_{pj} = (\mathbf{e}_p^{\mathrm{T}}\mathbf{B}^{-1})\mathbf{a}_j$, which can be calculated during the pricing operation (Step 2).

Harris (1973) has shown that this produces little error, and has produced some startling results which indicate at least a halving in the total number of iterations on large problems.

Since T_j is monotonically increasing, it may occasionally need resetting with $T_j = 1$, which corresponds to redefining the set of reference variables. The criterion for resetting is when the discrepancy between the approximate T_j and the exact value (obtained a posteriori once $\boldsymbol{\alpha}_q$ is calculated) exceeds a factor of, say, 2.

3-6-5 The Steepest-edge Pricing Method

Recently, Goldfarb and Reid (1977) have devised a scheme similar to Harris's in which the weightings to be applied to the reduced costs are maintained *exactly*, at little extra cost in computation. Let us suppose the qth nonbasic variable has been chosen to enter the basis. The new value of $\mathbf{x}$ can be expressed in the form

$$\bar{\mathbf{x}}_B = \boldsymbol{\beta} - \boldsymbol{\alpha}_q \theta \tag{3-47}$$

$$\bar{\mathbf{x}} = \left[\begin{array}{c} \bar{\mathbf{x}}_B \\ \hline \bar{\mathbf{x}}_N \end{array}\right] = \mathbf{x} + \left[\begin{array}{c} -\boldsymbol{\alpha}_q \\ \hline \mathbf{e}_q \end{array}\right] \theta \tag{3-48}$$

where $$\boldsymbol{\beta} = \mathbf{x}_B, \boldsymbol{\alpha}_q = \mathbf{B}^{-1}\mathbf{a}_q$$

(cf. Eqs. (2-18), (2-27)) and

$$\theta = \min_{i \mid \alpha_{iq} > 0} \left\{ \frac{\beta_i}{a_{iq}} \right\},$$

corresponding to the ratio test for the leaving basic variable.

Geometrically, we may regard the vector

$$\boldsymbol{\eta}_q = \left[\begin{array}{c} -\boldsymbol{\alpha}_q \\ \hline \mathbf{e}_q \end{array}\right] \tag{3-49}$$

as the "edge" of the polytope along which $\mathbf{x}$ moves from one vertex to another. The standard simplex pricing method involves choosing the most negative d_j: where

$$d_j = z_j - c_j = \mathbf{c}_B^{\mathrm{T}}\mathbf{B}^{-1}\mathbf{a}_j - \mathbf{c}_j = -\mathbf{c}^{\mathrm{T}}\boldsymbol{\eta}_j \qquad j = 1, \ldots, n - m \tag{3-50}$$

(cf. Eqs. (2-30)–(2-32).

In contrast, the *steepest-edge* pricing method chooses the most negative inner product of the objective row with the normalized edge vectors

$$-\mathbf{c}^{\mathrm{T}} \frac{\boldsymbol{\eta}_q}{\|\boldsymbol{\eta}_q\|} = \min_j \left\{ -\mathbf{c}^{\mathrm{T}} \frac{\boldsymbol{\eta}_j}{\|\boldsymbol{\eta}_j\|} \right\} \qquad j = 1, \ldots, n - m \tag{3-51}$$

As its name implies, this method corresponds to choosing the edge of steepest ascent in E^n of the objective function.

Goldfarb and Reid have developed a recursion for updating $\|\boldsymbol{\eta}_j\|$, $j = 1, \ldots, n - m$, from iteration to iteration, as it is evident that recalculating these norms for every nonbasic variable at each iteration takes prohibitively long for large problems.

We may regard E^n as being spanned by the vectors given by the columns of the following square, nonsingular matrix:

$$\mathbf{S} = \left[\begin{array}{c|c} \mathbf{B} & \mathbf{N} \\ \hline \mathbf{0} & \mathbf{I} \end{array}\right] \tag{3-52}$$

The rows of $\mathbf{S}$ correspond to the set of binding constraints

$$\mathbf{B}\mathbf{x}_B + \mathbf{N}\mathbf{x}_N = \mathbf{b}, \quad \text{and} \quad \mathbf{I}\mathbf{x}_N = \mathbf{0}$$

For comparison with Harris's Devex approach, we may regard her initial *reference space* as the space spanned by the last $n - m$ columns of $\mathbf{S}$.

The reader may easily verify that the inverse of $\mathbf{S}$ is given by

$$\mathbf{S}^{-1} = \left[\begin{array}{c|c} \mathbf{B}^{-1} & -\mathbf{B}^{-1}\mathbf{N} \\ \hline \mathbf{0} & \mathbf{I} \end{array}\right] \tag{3-53}$$

Note that $\boldsymbol{\eta}_j$, $j = 1, \ldots, n - m$, is given by the last $n - m$ columns of $\mathbf{S}^{-1}$.

Now when $\mathbf{a}_q$ replaces $\mathbf{B}\mathbf{e}_p$ in the basis, we have

$$\bar{\mathbf{S}} = [\mathbf{S} + \mathbf{e}_{m+q}(\mathbf{e}_p^\top - \mathbf{e}_{m+q}^\top)]\mathbf{P}_{p,m+q} \tag{3-54}$$

where $\mathbf{P}_{p,m+q}$ is a permutation matrix which interchanges columns p and $m + q$. The addition of $\mathbf{e}_{m+q}(\mathbf{e}_p^\top - \mathbf{e}_{m+q}^\top)$ to $\mathbf{S}$ has the effect of putting $\mathbf{e}_{m+q}$ into the pth column position, so that subsequent permutation by $\mathbf{P}_{p,m+q}$ restores the identity in the last $n - m$ columns, as well as interchanging $\mathbf{a}_q$ and $\mathbf{B}\mathbf{e}_p$. We thus have

$$\bar{\mathbf{S}} = \left[\begin{array}{c|c} \left[\begin{array}{c|c|c} & p & \\ \mathbf{B} & \mathbf{a}_q & \end{array}\right] & \left[\begin{array}{c|c|c} & m+q & \\ \mathbf{N} & \mathbf{B}\mathbf{e}_p & \end{array}\right] \\ \hline \mathbf{0} & \mathbf{I} \end{array}\right] \tag{3-55}$$

Using the Sherman–Morrison formula, Eq. (1-40), in Eq. (3-54), we get

$$\begin{aligned} \mathbf{P}_{p,m+q}\bar{\mathbf{S}}^{-1} &= \mathbf{S}^{-1} - \frac{\mathbf{S}^{-1}\mathbf{e}_{m+q}(\mathbf{e}_p - \mathbf{e}_{m+q})^\top\mathbf{S}^{-1}}{1 + (\mathbf{e}_p - \mathbf{e}_{m+q})^\top\mathbf{S}^{-1}\mathbf{e}_{m+q}} \\ &= \mathbf{S}^{-1} - \frac{\mathbf{S}^{-1}\mathbf{e}_{m+q}(\mathbf{e}_p^\top\mathbf{S}^{-1} - \mathbf{e}_{m+q}^\top)}{\mathbf{e}_p^\top\mathbf{S}^{-1}\mathbf{e}_{m+q}} \end{aligned} \tag{3-56}$$

since $\mathbf{e}_{m+q}^\top\mathbf{S}^{-1} = \mathbf{e}_{m+q}^\top$.

Thus we have

$$\mathbf{P}_{p,m+q}\bar{\boldsymbol{\eta}}_j = \mathbf{P}_{p,m+q}\bar{\mathbf{S}}^{-1}\mathbf{e}_{m+j} = \boldsymbol{\eta}_j - \boldsymbol{\eta}_q \cdot \left(\frac{\mathbf{e}_p^\top\boldsymbol{\eta}_j}{\mathbf{e}_p^\top\boldsymbol{\eta}_q}\right) \qquad j \neq q \tag{3-57}$$

$$\begin{aligned} \mathbf{P}_{p,m+q}\bar{\boldsymbol{\eta}}_q = \mathbf{P}_{p,m+q}\bar{\mathbf{S}}^{-1}\mathbf{e}_{m+q} &= \boldsymbol{\eta}_q - \boldsymbol{\eta}_q + \frac{\boldsymbol{\eta}_q}{\mathbf{e}_p^\top\boldsymbol{\eta}_q} \\ &= \boldsymbol{\eta}_q\left(\frac{1}{\mathbf{e}_p^\top\boldsymbol{\eta}_q}\right) \end{aligned} \tag{3-58}$$

Note that applying the permutation $\mathbf{P}_{p,m+q}$ puts the pth element of $\bar{\boldsymbol{\eta}}_q$, $\mathbf{e}_p^\top\boldsymbol{\eta}_q/\mathbf{e}_p^\top\boldsymbol{\eta}_q$ $(=1)$, in row position $m + q$, in column $m + q$ of $\mathbf{S}$, and likewise puts the pth element of $\bar{\boldsymbol{\eta}}_j$, $\mathbf{e}_p^\top\boldsymbol{\eta}_j - \mathbf{e}_p^\top\boldsymbol{\eta}_q\,(\mathbf{e}_p^\top\boldsymbol{\eta}_j/\mathbf{e}_p^\top\boldsymbol{\eta}_q)$ $(=0)$, in row position $m + q$, in other columns $m+j, j \neq q$, of $\mathbf{S}$, so that the identity is retained

$$\bar{\mathbf{S}}^{-1} = \left[\begin{array}{c|c} \bar{\mathbf{B}}^{-1} & -\bar{\mathbf{B}}^{-1}\bar{\mathbf{N}} \\ \hline \mathbf{0} & \mathbf{I} \end{array}\right] \tag{3-59}$$

Now $\mathbf{e}_p^{\mathsf{T}}\boldsymbol{\eta}_q = -\mathbf{e}_p^{\mathsf{T}}\mathbf{B}^{-1}\mathbf{a}_q = -\alpha_{pq}$, the pivot element, and, using Eq. (2-36),

$$\bar{\alpha}_{pj} = \mathbf{e}_p^{\mathsf{T}}\bar{\mathbf{B}}^{-1}\mathbf{a}_j = -\frac{\alpha_{pj}}{\alpha_{pq}} \qquad j = 1, \ldots, n-m, j \neq q \tag{3-60}$$

Thus the formula simplifies to

$$\mathbf{P}_{p,m+q}\bar{\boldsymbol{\eta}}_q = -\boldsymbol{\eta}_q/\alpha_{pq} \tag{3-61}$$

$$\mathbf{P}_{p,m+q}\bar{\boldsymbol{\eta}}_j = \boldsymbol{\eta}_j - \boldsymbol{\eta}_q\bar{\alpha}_{pj} \qquad j \neq q \tag{3-62}$$

Let

$$\gamma_j = \boldsymbol{\eta}_j^{\mathsf{T}}\boldsymbol{\eta}_j = \|\boldsymbol{\eta}_j\|^2 \qquad j = 1, \ldots, n-m \tag{3-63}$$

Then

$$\begin{aligned}\bar{\gamma}_q &= \gamma_q/\alpha_{pq}^2 \\ &= (1 + \boldsymbol{\alpha}_q^{\mathsf{T}}\boldsymbol{\alpha}_q)/\alpha_{pq}^2\end{aligned} \tag{3-64}$$

and

$$\begin{aligned}\bar{\gamma}_j &= (\boldsymbol{\eta}_j - \boldsymbol{\eta}_q\bar{\alpha}_{pj})^{\mathsf{T}}(\boldsymbol{\eta}_j - \boldsymbol{\eta}_q\bar{\alpha}_{pj}) \qquad j \neq q \\ &= \gamma_j - 2\bar{\alpha}_{pj}\boldsymbol{\eta}_j^{\mathsf{T}}\boldsymbol{\eta}_q + \bar{\alpha}_{pj}^2\gamma_q \\ &= \gamma_j - 2\bar{\alpha}_{pj}\mathbf{a}_j^{\mathsf{T}}(\mathbf{B}^{-\mathsf{T}}\boldsymbol{\alpha}_q) + \bar{\alpha}_{pj}^2\gamma_q\end{aligned} \tag{3-65}$$

Thus the extra computation required above that for the standard simplex algorithm is in the evaluation of $\mathbf{B}^{-\mathsf{T}}\boldsymbol{\alpha}_q$ and $\bar{\alpha}_{pj} = \mathbf{e}_p^{\mathsf{T}}\bar{\mathbf{B}}^{-1}\mathbf{a}_j, j = 1, \ldots, n-m$, and the only extra computation above Harris's algorithm is $\mathbf{B}^{-\mathsf{T}}\boldsymbol{\alpha}_q$ (and its inner product with $\mathbf{a}_j$).

Goldfarb and Reid have demonstrated that their algorithm generally requires fewer iterations than the Harris Devex method, and also compares favourably in terms of total run time. Both methods present a significant advance on the standard simplex method.

CHAPTER

FOUR

DUALITY AND POSTOPTIMALITY ANALYSIS

4-1 THE CANONICAL FORM

For each linear programming problem (*primal problem*) there exists a corresponding *dual problem* in an entirely different space of variables, but whose optimum value is the same (when a finite solution exists). The two problems are at their most symmetrical when presented in the so-called *canonical form*, as shown below.

Primal problem:

Maximize $$\mathbf{c}^{\mathsf{T}}\mathbf{x} \tag{4-1}$$

subject to $$\mathbf{A}\mathbf{x} \leqslant \mathbf{b} \tag{4-2}$$

$$\mathbf{x} \geqslant \mathbf{0} \tag{4-3}$$

Dual problem:

Minimize $$\mathbf{b}^{\mathsf{T}}\mathbf{y} \tag{4-4}$$

subject to $$\mathbf{A}^{\mathsf{T}}\mathbf{y} \geqslant \mathbf{c} \tag{4-5}$$

$$\mathbf{y} \geqslant \mathbf{0} \tag{4-6}$$

The symmetry of the two problems is self-evident; the dual of the dual problem is the primal problem. The right-hand side becomes the cost row and vice versa, and the rows of **A** (the constraints of the primal) become the columns of the dual. Each of the dual variables (**y**) is associated with a primal constraint.

Note that in the canonical form, the constraints of the primal, Eq. (4-2), are presented as inequalities; if they were equalities, the corresponding elements of **y** in the dual problem would be unrestricted in sign (i.e., Eq. (4-6) would not apply).

4-1-1 The Duality Theorem

If either problem has a finite optimal solution, then the other also has a finite optimal solution, and the two optima are equal in value.

If either has an unbounded solution, then the other has no feasible solution.

Consider the primal problem, Eqs. (4-1)–(4-3). Transforming the inequality constraints into equalities using slack variables $\mathbf{x}_s$ yields the enlarged problem

Maximize $$\mathbf{c}^\top \mathbf{x} + \mathbf{0}^\top \mathbf{x}_s \tag{4-7}$$

subject to $$\mathbf{A}\mathbf{x} + \mathbf{I}\mathbf{x}_s = \mathbf{b} \tag{4-8}$$

$$\mathbf{x}, \mathbf{x}_s \geqslant \mathbf{0} \tag{4-9}$$

Thus the enlarged problem has cost row $(\mathbf{c}^\top : \mathbf{0}^\top)$ and corresponding matrix $[\mathbf{A} : \mathbf{I}]$.

Let us *assume* that $\hat{\mathbf{x}}_B$ is the optimal basic feasible solution to the enlarged problem, $\mathbf{c}_B^\top$ is the corresponding partition of $(\mathbf{c}^\top : \mathbf{0}^\top)$, and $\mathbf{B}$ is the corresponding optimal basis matrix, so that $\hat{\mathbf{x}}_B = \mathbf{B}^{-1}\mathbf{b}$.

Under this assumption, the reduced cost d_j of all nonbasic variables is nonnegative,

$$d_j = \mathbf{c}_B^\top \mathbf{B}^{-1}\mathbf{a}_j - c_j \geqslant 0 \qquad \text{all } j \text{ nonbasic} \tag{4-10}$$

Rearranging, $$\mathbf{c}_B^\top \mathbf{B}^{-1}\mathbf{a}_j \geqslant c_j \qquad \text{all } j \text{ nonbasic} \tag{4-11}$$

Note also that the reduced cost of basic variables is zero, since

$$\mathbf{c}_B^\top \mathbf{B}^{-1}\mathbf{B} - \mathbf{c}_B^\top = \mathbf{0}^\top, \text{ i.e., } \mathbf{c}_B^\top \mathbf{B}^{-1}\mathbf{a}_j = c_j \qquad \text{all } j \text{ basic} \tag{4-12}$$

Since a variable is either basic or nonbasic, each $\mathbf{a}_j$ is a column of $[\mathbf{A} : \mathbf{I}]$, and c_j is an element of $[\mathbf{c}^\top : \mathbf{0}^\top]$, and we have, considering the whole matrix $[\mathbf{A} : \mathbf{I}]$,

$$\mathbf{c}_B^\top \mathbf{B}^{-1}[\mathbf{A} : \mathbf{I}] \geqslant (\mathbf{c}^\top : \mathbf{0}^\top) \tag{4-13}$$

Let us *define* $\hat{\mathbf{y}}$ by the equation

$$\hat{\mathbf{y}}^\top = \mathbf{c}_B^\top \mathbf{B}^{-1} \tag{4-14}$$

Then, using Eq. (4-14) in Eq. (4-13) and taking each partition in turn,

$$\hat{\mathbf{y}}^\top \mathbf{A} \geqslant \mathbf{c}^\top \qquad \text{i.e.} \qquad \mathbf{A}^\top \hat{\mathbf{y}} \geqslant \mathbf{c} \tag{4-15}$$

$$\hat{\mathbf{y}}^\top \mathbf{I} \geqslant \mathbf{0}^\top \qquad \text{i.e.} \qquad \hat{\mathbf{y}} \geqslant \mathbf{0} \tag{4-16}$$

But Eqs. (4-15) and (4-16) correspond to Eqs. (4-5) and (4-6). We thus have,

Proposition 1 An optimal solution to the primal provides a feasible solution to the dual (and vice versa).

Note that this result holds regardless of whether $\hat{\mathbf{x}}_B$ is feasible, for we nowhere used the feasibility of $\hat{\mathbf{x}}_B$ in arriving at Eqs. (4-15) and (4-16).

Consider the cost function of the dual problem, Eq. (4-4). Using Eq. (4-14),

$$\mathbf{b}^{\top}\hat{\mathbf{y}} = \hat{\mathbf{y}}^{\top}\mathbf{b} = \mathbf{c}_B^{\top}\mathbf{B}^{-1}\mathbf{b} = \mathbf{c}_B^{\top}\hat{\mathbf{x}}_B \tag{4-17}$$

But for any feasible solution to the primal $\mathbf{x}$, and any feasible solution to the dual $\mathbf{y}$, Eqs. (4-2), (4-3), (4-5), and (4-6) hold. From Eq. (4-2),

$$\mathbf{Ax} \leqslant \mathbf{b} \tag{4-18}$$

so

$$\mathbf{y}^{\top}\mathbf{Ax} \leqslant \mathbf{y}^{\top}\mathbf{b} \qquad \text{since } \mathbf{y} \geqslant \mathbf{0} \tag{4-19}$$

From Eq. (4-5),

$$\mathbf{y}^{\top}\mathbf{A} \geqslant \mathbf{c}^{\top} \tag{4-20}$$

so

$$\mathbf{y}^{\top}\mathbf{Ax} \geqslant \mathbf{c}^{\top}\mathbf{x} \qquad \text{since } \mathbf{x} \geqslant \mathbf{0} \tag{4-21}$$

Thus

$$\mathbf{y}^{\top}\mathbf{b} \geqslant \mathbf{c}^{\top}\mathbf{x} \tag{4-22}$$

for all feasible $\mathbf{x}$ and $\mathbf{y}$.

However, for the particular value $\hat{\mathbf{y}} = \mathbf{c}_B^{\top}\mathbf{B}^{-1}$, we have from Eqs. (4-17) and (4-22),

$$\mathbf{y}^{\top}\mathbf{b} \geqslant \mathbf{c}_B^{\top}\hat{\mathbf{x}}_B = \hat{\mathbf{y}}^{\top}\mathbf{b} \tag{4-23}$$

so $\hat{\mathbf{y}}$ is optimal, since it minimizes $\mathbf{y}^{\top}\mathbf{b}$ for all feasible $\mathbf{y}$.

Proposition 2 If $\hat{\mathbf{x}}$ is a feasible solution to the primal, and $\hat{\mathbf{y}}$ is a feasible solution to the dual, and

$$\mathbf{c}^{\top}\hat{\mathbf{x}} = \mathbf{b}^{\top}\hat{\mathbf{y}} \tag{4-24}$$

then $\hat{\mathbf{x}}$ is an optimal solution to the primal, and $\hat{\mathbf{y}}$ is an optimal solution to the dual.

If the primal is unbounded, then from Eq. (4-22),

$$\mathbf{y}^{\top}\mathbf{b} \geqslant \infty \tag{4-25}$$

Any solution to Eqs. (4-5)–(4-6) must have a corresponding value for the dual objective function which is an upper bound for the primal objective function. But this contradicts the assumption of unboundedness, so we conclude there is no feasible solution to the dual problem.

Consider the slack variables associated with the primal constraints. Their reduced costs are given by the vector

$$(\mathbf{c}_B^{\top}\mathbf{B}^{-1}\mathbf{I} - \mathbf{0}^{\top}) = \mathbf{c}_B^{\top}\mathbf{B}^{-1} = \hat{\mathbf{y}}^{\top} \tag{4-26}$$

Thus the optimal values of the dual variables $\hat{\mathbf{y}}$ are given by the reduced costs of the slack variables of the primal constraints at the optimum.

To see why equality constraints give rise to dual variables unrestricted in sign, we could replace the equality constraints

$$\mathbf{Ax} = \mathbf{b} \tag{4-27}$$

by the two inequalities

$$\mathbf{Ax} \leqslant \mathbf{b} \tag{4-28}$$

and

$$\mathbf{Ax} \geqslant \mathbf{b} \quad \text{i.e.} \quad -\mathbf{Ax} \leqslant -\mathbf{b} \tag{4-29}$$

Thus the dual problem corresponding to this becomes

Minimize

$$\mathbf{b}^{\top}\mathbf{y}_1 - \mathbf{b}^{\top}\mathbf{y}_2 \tag{4-30}$$

subject to

$$\mathbf{A}^{\top}\mathbf{y}_1 - \mathbf{A}^{\top}\mathbf{y}_2 \geqslant \mathbf{c} \tag{4-31}$$

$$\mathbf{y}_1, \mathbf{y}_2 \geqslant \mathbf{0}$$

i.e., Minimize

$$\mathbf{b}^{\top}\mathbf{y} \tag{4-32}$$

subject to

$$\mathbf{A}^{\top}\mathbf{y} \geqslant \mathbf{c} \tag{4-33}$$

where

$$\mathbf{y} = \mathbf{y}_1 - \mathbf{y}_2 \tag{4-34}$$

Although $\mathbf{y}_1$ and $\mathbf{y}_2$ are each constrained to be nonnegative, the difference of two nonnegative numbers can take any value, so $\mathbf{y}$ is unrestricted in sign.

4-2 SHADOW PRICES: ECONOMIC INTERPRETATION

The canonical form also provides us with an economic interpretation of the values of the dual variables. Recall from the previous section that at the optimum the dual variables $\hat{\mathbf{y}}$ are given by the reduced costs of the slack variables associated with the primal constraints.

Suppose the slack variable $(\mathbf{x}_s)_i$ of constraint i is nonbasic at the optimum, where constraint i is expressed in the form

$$\sum_{j=1}^{n} a_{ij}x_j + (\mathbf{x}_s)_i = b_i \tag{4-35}$$

Since $(\mathbf{x}_s)_i$ is nonbasic at zero, the original constraint $\sum_{j=1}^{n} a_{ij}x_j \leqslant b_i$ is active as a strict equality at the optimum, i.e.,

$$\sum_{j=1}^{n} a_{ij}\hat{x}_j = b_i \tag{4-36}$$

Now, by definition, the reduced cost of this nonbasic variable is the amount by which the objective function P could increase per unit increase in this variable. Since the solution is optimal, the reduced cost is positive (nonnegative), so the objective function must decrease if the slack variable is increased by 1 unit, and increase if the slack variable is decreased by 1 unit. Suppose the right-hand side term b_i was increased by 1 unit, so that the constraint becomes

$$\sum_{j=1}^{n} a_{ij}\hat{x}_j = b_i + 1 \tag{4-37}$$

or rearranging,

$$\sum_{j=1}^{n} a_{ij}\hat{x}_j + (-1) = b_i \tag{4-38}$$

that is, the slack variable $(\mathbf{x}_s)_i$ would take a value of -1 for the ith constraint to remain active as an equality, and the reduced cost gives the corresponding increase in objective function value.

Thus, the reduced cost of the ith slack variable gives the *increase* in objective function value per unit increase in right-hand side coefficient b_i. Since the b_i term usually represents a bound on resource availability the reduced cost is called the *shadow price* of that constraint, since it represents the value of an extra unit of that resource. It is a marginal price in that it is valid only over the range in which the current basis remains optimal. (We cannot actually have $(\mathbf{x}_s)_i$ at -1 in the tableau, so for $(\mathbf{x}_s)_i$ to remain nonbasic at zero we must alter the values of our current basic variables $\hat{\mathbf{x}}_B$, which in turn must remain nonnegative. Determination of the range is discussed in Sec. 4-4.)

The same result can be obtained in a quicker but somewhat less meaningful way by considering Eq. (4-17). At the optimum, the objective function is given by

$$P = \mathbf{c}_B^\top \hat{\mathbf{x}}_B = \hat{\mathbf{y}}^\top \mathbf{b} = \sum_{i=1}^{m} \hat{y}_i b_i \tag{4-39}$$

Thus the incremental change in the objective function P per unit change in b_i is given by

$$\frac{\partial P}{\partial b_i} = y_i \qquad i = 1, \ldots, m \tag{4-40}$$

which is the reduced cost of the ith slack variable.

This remains valid over the range in which $\hat{\mathbf{y}} = \mathbf{c}_B^\top \mathbf{B}^{-1}$ remains constant, which is the range in which $\mathbf{B}$ remains the optimal basis.

If a slack variable is basic at the optimal solution, then by definition its reduced cost is zero. But this is also meaningful since if a resource is not fully utilized $(\sum_{j=1}^{n} a_{ij}\hat{x}_j < b_i)$, then the price we would pay for an extra unit of that resource is zero. This gives rise to the *principle of complementary slackness*:

At an optimal solution

$$\text{either} \qquad \sum_{j=1}^{n} a_{ij}\hat{x}_j = b_i \quad \text{or} \quad \hat{y}_i = 0 \text{ (or both)}, \; i = 1, \ldots, m \tag{4-41}$$

Similarly, considering the dual,

$$\text{either} \qquad \sum_{i=1}^{m} a_{ij}\hat{y}_i = c_j \quad \text{or} \quad \hat{x}_j = 0 \text{ (or both)}, \; j = 1, \ldots, n \tag{4-42}$$

The dual variables can be obtained more readily in practice than by looking at the reduced cost of the slack variables $(\hat{\mathbf{y}} = \mathbf{c}_B^\top \mathbf{B}^{-1}\mathbf{I} - \mathbf{0}^\top)$. Note in Step 1 of the simplex method (Sec. 2-6) we produced the pricing vector

$$\boldsymbol{\pi}^\top = \mathbf{c}_B^\top \mathbf{B}^{-1} \tag{4-43}$$

Thus,
$$\hat{\mathbf{y}}^\top = \boldsymbol{\pi}^\top \quad \text{at the optimum.} \tag{4-44}$$

Since the pricing vector $\boldsymbol{\pi}^\top$ is computed at every iteration of the simplex method (as well as at the optimum) we have at any iteration

$$P = \mathbf{c}_B^\top \mathbf{x}_B = \mathbf{c}_B^\top \mathbf{B}^{-1}\mathbf{b} = \boldsymbol{\pi}^\top \mathbf{b} \tag{4-45}$$

Thus the pricing vector $\boldsymbol{\pi}^\top$ gives the rate of change of objective function with respect to the right-hand side vector **b** at any iteration; this fact will be put to good use when discussing decomposition methods in Chapter 5.

Note that if we *define* our dual variables **y** by the equation

$$\mathbf{y}^\top = \boldsymbol{\pi}^\top = \mathbf{c}_B^\top \mathbf{B}^{-1} \tag{4-46}$$

for *each* iteration of the simplex method, then the inequality (4-22) is always satisfied as an equality, but **y** is infeasible throughout the course of the simplex iterations until optimality is reached.

4-3 PRICE MARGINS

The shadow prices are associated with constraints rather than variables. However, they are often used to evaluate prices or cost coefficients associated with the variables of the primal problem.

As an example, suppose we have an **A** matrix which represents the daily operation of an oil refinery, and a particular variable x_j representing the purchase of crude oil feedstock, with a cost of 12.65 \$/barrel ($c_j = -12.65$).

There is an upper limit on the purchase of this oil of 50 M barrels/day† at this price. This is represented by the constraint

$$x_j + x_s = 50 \tag{4-47}$$

where x_s is the associated slack variable. At the optimum, x_s has a reduced cost of 1.04 \$/barrel: what does this mean? The shadow price on that constraint is 1.04 \$/barrel, but this does *not* mean that we should only pay \$1.04 for another barrel of that crude. It means we should be prepared to pay another 1.04 \$/barrel for an opportunity to purchase extra supplies *given* that any further purchases would cost 12.65 \$/barrel; i.e., the objective function will increase by \$1.04 for every extra barrel we can purchase at the price c_j already in the cost row.

This means we should be prepared to bid up to 12.65 + 1.04 = 13.69 \$/barrel on the spot-market for extral supplies of that crude.

Note that 13.69 \$/barrel is the *breakeven price*, in that we increase our objective function P if we can purchase for less than this price, decrease P if we

† M = thousand, MM = million.

purchase for more, and make no change at all to P if we purchase for exactly 13.69 \$/barrel.

If we define

$$\begin{aligned} \text{Price margin} &= \text{breakeven price} - \text{actual price} \\ \text{then this, in our example} &= 13.69 - 12.65 \\ &= 1.04 \text{ \$/barrel} \end{aligned} \tag{4-48}$$

The price margin of x_j is the net profit made from each unit of x_j which can be purchased in excess of the existing upper bound, and is equal to the shadow price of the bounding constraint. It is valid only within a certain range of the existing optimum, corresponding to the range within which the current basis remains optimal, on both the upward side (increasing purchases) and downward side (decreasing purchases – with a corresponding net loss) of the current bound. Calculation of the range is discussed in Sec. 4-4.

An alternative way of calculating price margins is discussed in Sec. 4-6-3. In Chapter 5 we shall see that simple upper bounds can be dealt with implicitly by allowing a variable to be nonbasic at its *upper* bound, so that constraints of the form of Eq. (4-47) are not required. However, the same information is available as the reduced cost of the variable; in our example, x_j would be nonbasic at its upper bound of 50 and have a reduced cost of -1.04 \$/barrel.

The reduced cost of a variable nonbasic at its *lower* bound is often referred to as the *opportunity cost* of that variable. If management made the (nonoptimal) decision of increasing that nonbasic from its lower bound, the reduced cost gives the decrease in P per unit increase in the variable (for a certain range). This represents the opportunity loss in departing from the optimal solution.

Because they indicate how the values of the current basic variables change, the terms of $\boldsymbol{\alpha}_j$, j nonbasic, are often referred to in this context as the *marginal rates of substitution*, so that α_{ij} is the marginal rate of substitution of activity j for the activity i (cf. Eq. (2-28)).

4-4 RANGING

It is often said that postoptimality analysis is the most important part of LP calculations, and it is not hard to see why this conclusion is reached. Most of the coefficients which appear in an LP problem are not known exactly, and in practice are usually best estimates of the value that the coefficient should be. We are thus interested in the range of variation of these coefficients for which the optimal solution remains optimal, in the sense that the basis does not change.

Three categories are investigated here: cost coefficients c_j, right-hand side terms b_i, and matrix coefficients a_{ij}.

4-4-1 Changes in the Cost Row

(a) Nonbasic variable The change in the cost coefficient of a nonbasic variable affects the reduced cost of that variable only, and the change is in direct proportion,

$$\bar{c}_j = c_j + \delta \Rightarrow \bar{d}_j = d_j - \delta \tag{4-49}$$

Recall from Eq. (2-22) that the reduced cost is given by

$$d_j = \boldsymbol{\pi}^\mathsf{T}\mathbf{a}_j - c_j \tag{4-50}$$

Thus

$$\bar{d}_j = \boldsymbol{\pi}^\mathsf{T}\mathbf{a}_j - \bar{c}_j \tag{4-51}$$

$$= \boldsymbol{\pi}^\mathsf{T}\mathbf{a}_j - (c_j + \delta) \tag{4-52}$$

$$= d_j - \delta \tag{4-53}$$

which is as given in Eq. (4-49).

Example 1 Suppose the **A** matrix represents an industrial process, and a particular variable x_j represents the amount of a particular product produced, with selling price $c_j = \$20$/unit. At the optimum solution it is nonbasic $(=0)$ with a reduced cost of \$1.40/unit. Thus, if the price increases to \$21.40/unit, the reduced cost will be zero, and any further increase in price will give a negative reduced cost. This means that the current solution will then no longer be optimal, in that it will then be profitable to produce the product represented by x_j. Thus \$21.40/unit is the *breakeven price* of x_j; at any lower price, it is optimal to produce none (x_j remains nonbasic), and at any higher price it is more profitable for x_j to become basic.

For a nonbasic variable, the *range* over which c_j can vary and the current solution remain optimal is given by $c_j + \delta$, where

$$-\infty < \delta \leqslant \hat{d}_j \tag{4-54}$$

where $\hat{d}_j$ is the reduced cost at the current optimum.

Note that for any arbitrary *negative* δ, the reduced cost will remain positive.

It is worth mentioning that in most commercial LP codes, there is another range given at the same time as well, and that is the range that x_j can be increased from zero before a change of basis occurs. When $\delta = \hat{d}_j$, the reduced cost is zero, which implies that x_j can be increased without affecting the value of the objective function. The range that it can increase is given by $\min_{i|\alpha_{ij}>0} (\boldsymbol{\beta}/\boldsymbol{\alpha}_j)_i$, where $\boldsymbol{\alpha}_j = \mathbf{B}^{-1}\mathbf{a}_j$, as in Step 4 of the simplex method (Sec. 2-6).

Example 2 Considering Example 1 further, suppose that at the optimum the vector of basic variables, current right-hand side $\boldsymbol{\beta} = \mathbf{B}^{-1}\mathbf{b}$, and $\boldsymbol{\alpha}_j = \mathbf{B}^{-1}\mathbf{a}_j$ are given by

$$\mathbf{x}_B = \begin{bmatrix} x_5 \\ x_1 \\ x_6 \end{bmatrix} \qquad \boldsymbol{\beta} = \begin{bmatrix} 3.2 \\ 1.5 \\ 5.6 \end{bmatrix} \qquad \boldsymbol{\alpha}_j = \begin{bmatrix} 0.6 \\ 0.3 \\ -1.2 \end{bmatrix} \tag{4-55}$$

We have

$$\min_{i|\alpha_{ij}>0}\left(\frac{\beta_i}{\alpha_{ij}}\right)=\frac{1.5}{0.3}=5.0$$

We thus conclude that at a selling price of \$21.40/unit or more it becomes profitable to produce x_j; for each unit produced, x_5, x_1, and x_6 decrease by 0.6, 0.3, and -1.2 units respectively; and we can produce 5.0 units of x_j before x_1 becomes zero and a change of basis is required. Note that we have got all this information *without having to re-solve the problem*, but to pursue the analysis any further requires a pivot step corresponding to the change of basis.

(b) Basic variable A change in the cost coefficient of a basic variable affects the reduced cost of the nonbasic variables. Suppose we incremented the cost coefficient of the ith basic variable. This affects the vector as follows:

$$\bar{\mathbf{c}}_B=\mathbf{c}_B+\delta\mathbf{e}_i \tag{4-56}$$

The reduced cost of the jth nonbasic variable is now

$$\bar{d}_j=(\mathbf{c}_B+\delta\mathbf{e}_i)^{\mathsf{T}}\boldsymbol{\alpha}_j-c_j \tag{4-57}$$

$$=(\mathbf{c}_B^{\mathsf{T}}\boldsymbol{\alpha}_j-c_j)+\delta\mathbf{e}_i^{\mathsf{T}}\boldsymbol{\alpha}_j \tag{4-58}$$

$$=d_j+\delta\alpha_{ij} \tag{4-59}$$

where $\alpha_{ij}=(\mathbf{B}^{-1}\mathbf{a}_j)_i$ is computed for all nonbasic variables by solving $\mathbf{B}^{\mathsf{T}}\mathbf{y}=\mathbf{e}_i$ for $\mathbf{y}$, then computing $\alpha_{ij}=\mathbf{y}^{\mathsf{T}}\mathbf{a}_j$ for each nonbasic. (Obviously if $\alpha_{ij}=0$ for any variable, the reduced cost does not change.)

For a solution to remain optimal, we must have $\bar{d}_j\geqslant 0$, i.e.,

$$\hat{d}_j+\delta\alpha_{ij}\geqslant 0 \tag{4-60}$$

where $\hat{d}_j$ is the reduced cost at the current optimum.

For a basic variable, the *range* over which c_i can vary and the current solution remain optimal is given by $c_i+\delta$, where

$$\max_{j|\alpha_{ij}>0}\left\{\frac{\hat{d}_j}{-\alpha_{ij}}\right\}\leqslant\delta\leqslant\min_{j|\alpha_{ij}<0}\left\{\frac{\hat{d}_j}{-\alpha_{ij}}\right\} \tag{4-61}$$

since this is the range for which $\hat{d}_j+\delta\alpha_{ij}\geqslant 0$.

If there is no $\alpha_{ij}<0$, then $\delta<\infty$, and likewise if there is no $\alpha_{ij}>0$, then $\delta>-\infty$.

Example 3 Suppose we had an optimal tableau of the form

Maximize $\quad P=31.5-3.5x_4-0.1x_3-0.25x_5$

subject to

$$\begin{aligned} x_1&=3.2-1.0x_4-0.5x_3-0.6x_5\\ x_2&=1.5+0.5x_4+1.0x_3-1.0x_5\\ x_6&=5.6-2.0x_4-0.5x_3-1.0x_5 \end{aligned} \tag{4-62}$$

If the cost coefficient of x_2 becomes $c_2 + \delta$, the reduced cost of the nonbasic variables are changed as follows:

$$x_4 : \bar{d}_j = 3.5 + \delta(-0.5)$$
$$x_3 : \bar{d}_j = 0.1 + \delta(-1.0)$$
$$x_5 : \bar{d}_j = 0.25 + \delta(+1.0)$$

(Note that $\mathbf{x}_B = \boldsymbol{\beta} - \boldsymbol{\alpha}_j x_j$, so that α_{ij} is the negative of the number appearing in the tableau, Eq. (4-62)).

The range that δ can take is given by

$$\left(\frac{0.25}{-1.0}\right) \leqslant \delta \leqslant \min\left(\frac{3.5}{0.5}, \frac{0.1}{1.0}\right) \tag{4-63}$$

i.e.,

$$-0.25 \leqslant \delta \leqslant 0.1 \tag{4-64}$$

When δ is taken to the limit of its range, so that a reduced cost does become zero, the range to which the corresponding nonbasic variable can increase is calculated the same way as in Example 2.

In this example, for $\delta = 0.1$ the reduced cost of x_3 is zero, so that if the cost coefficient of x_2 increases by 0.1 or beyond it becomes profitable to produce x_3, and we can produce $\min\{3.2/0.5, 5.6/0.5\} = 6.4$ units of x_3 before x_1 becomes zero and a change of basis is required.

To summarize:

1. There is a range of price changes δ for both basic and nonbasic variables for which the current optimal solution remains optimal. For nonbasic variables there is an upper limit on δ; for basic variables there is usually both upper and lower limits on δ. Changing a price beyond these limits makes the original optimal solution nonoptimal, since the reduced cost of a nonbasic variable becomes negative.
2. Changing the price of a basic variable changes the value of the objective function, since $P = \mathbf{c}_B^T \boldsymbol{\beta}$ and the vector $\mathbf{c}_B$ is changed.
3. There are two possible viewpoints from which to consider the effect of price changes: from a marketing viewpoint we are interested in breakeven prices; from a production viewpoint we are interested in the range of price changes over which the current operating plan (represented by the current basis) remains optimal.

4-4-2 Changes in the Right-hand Side Vector

We wish to consider the effect of a change

$$\bar{b}_i = b_i + \delta \qquad \text{for some } 1 \leqslant i \leqslant m \tag{4-65}$$

It is usual to consider the case where b_i is the right-hand side of an *inequality* constraint, which therefore has a slack variable associated with it. We wish to determine the range over which the current optimal feasible solution remains so. We could analyze an equality constraint by regarding its associated artificial as a positive slack (which must be nonbasic for a feasible solution).

(a) Basic slack variable If the slack variable associated with the ith constraint is basic, then that constraint is not binding at the optimum. The analysis is simple: the value of the basic slack variable gives the range over which the right-hand side b_i can be reduced (increased for $\geqslant$ constraint).

The solution remains feasible and optimal for the range $b_i + \delta$, where

$$\begin{aligned} -\hat{x}_s \leqslant \delta < \infty \qquad & \text{for } \leqslant \text{ type constraint} \\ -\infty < \delta \leqslant \hat{x}_s \qquad & \text{for } \geqslant \text{ type constraint} \end{aligned} \tag{4-66}$$

where $\hat{x}_s$ is the value of the associated slack variable.

Example 4 Consider the inequality constraint

$$3x_1 + 4x_2 + 7x_3 \leqslant 100 \tag{4-67}$$

With, say, x_8 as the associated slack variable this becomes

$$3x_1 + 4x_2 + 7x_3 + x_8 = 100 \tag{4-68}$$

If, at the optimum, $x_8 = 26$, then the variables satisfy

$$3x_1 + 4x_2 + 7x_3 \leqslant 74 \tag{4-69}$$

as well as any higher right-hand side value.

(b) Nonbasic slack variable If a slack variable is nonbasic at zero, then the original inequality constraint is binding at the optimum. At first sight it would seem that because the constraint is binding, there is no possibility of changing the right-hand side term, particularly in *decreasing* the value of b_i (for $\leqslant$ type constraints). It turns out that by changing the vector $\mathbf{b}$ we also change $\mathbf{x}_B(=\mathbf{B}^{-1}\mathbf{b})$ and there is a range over which $\mathbf{x}_B$ remains nonnegative, so that we still retain an optimal feasible solution, in the sense that the basis does not change. (Note that both $\mathbf{x}_B$ and $P = \mathbf{c}_B^T\mathbf{x}_B$ change value.)

Consider the constraint

$$a_{k1}x_1 + a_{k2}x_2 + \cdots + x_s = b_k \tag{4-70}$$

where x_s is the slack variable. If the right-hand side becomes $b_k + \delta$, Eq. (4-70) is rearranged to become

$$a_{k1}x_1 + a_{k2}x_2 + \cdots + (x_s - \delta) = b_k \tag{4-71}$$

so that $(x_s - \delta)$ replaces x_s.

Thus if x_s is nonbasic at zero in the final tableau, we have the expression

$$\mathbf{x}_B = \boldsymbol{\beta} - \boldsymbol{\alpha}_s(-\delta) \tag{4-72}$$

where $\boldsymbol{\alpha}_s$ is the column in the tableau corresponding to x_s.

Since $\mathbf{x}_B$ must remain nonnegative, we have

$$\boldsymbol{\beta} - \boldsymbol{\alpha}_s(-\delta) \geqslant 0 \tag{4-73}$$

which is used to solve for the range over which δ can vary,

$$\max_{i|\alpha_{is}>0} \left\{\frac{\beta_i}{-\alpha_{is}}\right\} \leqslant \delta \leqslant \min_{i|\alpha_{is}<0} \left\{\frac{\beta_i}{-\alpha_{is}}\right\} \tag{4-74}$$

If there is no $\alpha_{is} > 0$, then $\delta > -\infty$, and if there is no $\alpha_{is} < 0$, then $\delta < \infty$.

For $\geqslant$ type constraints, δ changes sign, for we can analyze $\sum_{j=1}^{n} a_{ij}x_j \geqslant b_i$ in the form $-\sum a_{ij}x_j \leqslant -b_i$, so that $-(x_s + \delta)$ replaces $+(x_s - \delta)$ in Eq. (4-71).

Another way of seeing this is to consider the change to the right-hand side in the form

$$\bar{\mathbf{b}} = \mathbf{b} + \delta\mathbf{e}_k \tag{4-75}$$

Thus the new value of $\mathbf{x}_B$, $\bar{\mathbf{x}}_B$, is given by

$$\begin{aligned} \bar{\mathbf{x}}_B &= \mathbf{B}^{-1}\bar{\mathbf{b}} = \mathbf{B}^{-1}\mathbf{b} + \delta\mathbf{B}^{-1}\mathbf{e}_k \\ &= \boldsymbol{\beta} + \delta\mathbf{B}^{-1}\mathbf{e}_k \end{aligned} \tag{4-76}$$

But $\qquad \boldsymbol{\alpha}_s = \mathbf{B}^{-1}\mathbf{e}_k \qquad$ for a $\leqslant$ type constraint

and $\qquad \boldsymbol{\alpha}_s = -\mathbf{B}^{-1}\mathbf{e}_k \qquad$ for a $\geqslant$ type constraint

since the column corresponding to the slack variable is $+\mathbf{e}_k$ for a less than or equal ($\leqslant$) constraint and $-\mathbf{e}_k$ for a greater than or equal ($\geqslant$) constraint.

Thus we have

$$\text{for a} \leqslant \text{type constraint, } \boldsymbol{\beta} - \boldsymbol{\alpha}_s(-\delta) \geqslant \mathbf{0} \tag{4-77}$$

$$\text{for a} \geqslant \text{type constraint, } \boldsymbol{\beta} - \boldsymbol{\alpha}_s(+\delta) \geqslant \mathbf{0} \tag{4-78}$$

Example 5 Consider Example 3 again, and suppose x_4 represents a slack variable for a particular constraint i ($\leqslant$ type). If the coefficient b_i is varied by an amount δ, we have

$$\begin{aligned} x_1 &= 3.2 - 1.0(-\delta) \\ x_2 &= 1.5 + 0.5(-\delta) \\ x_6 &= 5.6 - 2.0(-\delta) \end{aligned} \qquad \text{i.e., } \boldsymbol{\beta} = \begin{bmatrix} 3.2 \\ 1.5 \\ 5.6 \end{bmatrix} \qquad \boldsymbol{\alpha}_s = \begin{bmatrix} 1.0 \\ -0.5 \\ 2.0 \end{bmatrix}$$

Thus

$$x_1 \geqslant 0 \quad \text{for} \quad 3.2 - 1.0(-\delta) \geqslant 0 \qquad \text{i.e. } \delta \geqslant \frac{3.2}{-1.0}$$

$$x_2 \geqslant 0 \quad \text{for} \quad 1.5 + 0.5(-\delta) \geqslant 0 \qquad \text{i.e. } \delta \leqslant \frac{1.5}{0.5}$$

$$x_6 \geqslant 0 \quad \text{for} \quad 5.6 - 2.0(-\delta) \geqslant 0 \qquad \text{i.e. } \delta \geqslant \frac{5.6}{-2.0}$$

δ can therefore vary in the range

$$\max\left\{\frac{3.2}{-1.0}, \frac{5.6}{-2.0}\right\} \leqslant \delta \leqslant \frac{1.5}{0.5}$$

i.e.,

$$-2.8 \leqslant \delta \leqslant 3.0$$

4-4-3 Changes in Matrix Coefficients

The coefficients a_{ij} are usually known with much more certainty than the cost row or right-hand side vector, since they usually represent some physical interaction between variables and are not subject to the same market fluctuations as prices and availabilities.

We shall consider changes to the coefficients of nonbasic variables only; changes to coefficients of basic variables alters the basis matrix $\mathbf{B}$ and are rather complicated to analyze (see, for example, Gal, 1979).

Consider the jth nonbasic variable, with corresponding column $\mathbf{a}_j$. If the ith element of $\mathbf{a}_j$ is changed by an amount δ, this affects the reduced cost d_j, as follows.

If

$$\bar{\mathbf{a}}_j = \mathbf{a}_j + \delta \mathbf{e}_i \tag{4-79}$$

then

$$\begin{aligned} \bar{d}_j &= \boldsymbol{\pi}^{\mathsf{T}}(\mathbf{a}_j + \delta \mathbf{e}_i) - c_j \\ &= d_j + \delta \boldsymbol{\pi}^{\mathsf{T}} \mathbf{e}_i \\ &= d_j + \delta \pi_i \end{aligned} \tag{4-80}$$

where $\boldsymbol{\pi}^{\mathsf{T}}(=\mathbf{c}_B^{\mathsf{T}}\mathbf{B}^{-1})$ is the pricing vector.

Thus the solution remains optimal ($\bar{d}_j \geqslant 0$) in the range

$$\begin{aligned} \delta &\leqslant \frac{d_j}{-\pi_i} \qquad \pi_i < 0 \\ \delta &\geqslant \frac{-d_j}{\pi_i} \qquad \pi_i > 0 \end{aligned} \tag{4-81}$$

4-5 DEGENERACY

4-5-1 Primal Degeneracy

Primal degeneracy occurs when a basic variable has a value of zero, i.e., one of the elements of $\boldsymbol{\beta} = \mathbf{B}^{-1}\mathbf{b}$ is zero. Primal degeneracy can often occur during intermediate (nonoptimal) basic feasible solutions. Thus no progress is made if the corresponding element of $\boldsymbol{\alpha}_q$ for the entering nonbasic variable is positive.

Suppose the ith element of $\boldsymbol{\beta}$ is zero. The tableau looks like

$$\begin{bmatrix} \\ \\ \\ \mathbf{x}_B \\ \\ \\ \\ \end{bmatrix} = \begin{bmatrix} \beta_1 \\ \beta_2 \\ \vdots \\ 0 \\ \vdots \\ \beta_m \end{bmatrix} - \begin{bmatrix} \alpha_{1q} \\ \alpha_{2q} \\ \vdots \\ \alpha_{iq} \\ \vdots \\ \alpha_{mq} \end{bmatrix} (\mathbf{x}_N)_q \tag{4-82}$$

If α_{ij} is positive, then we *must* pivot on $(\mathbf{x}_B)_i$; but $(\mathbf{x}_N)_q$ takes on a value $0/\alpha_{iq} = 0$, so becomes basic at zero and no progress is made, P does not increase, and we have merely interchanged $(\mathbf{x}_N)_q$ with $(\mathbf{x}_B)_i$ to be basic at zero. As mentioned in Sec. 2-4, this situation can give rise to *cycling*, in that a number of pivot operations are performed which eventually arrive back at the same basis.

It is possible to have degenerate intermediate solutions without the final optimal solution being so. If the primal optimal solution is degenerate, then there are infinitely many optimal solutions to the dual.

4-5-2 Dual Degeneracy

Dual degeneracy occurs when the reduced cost of a nonbasic variable is zero, i.e., one of the elements of the vector $(\boldsymbol{\pi}^\mathsf{T}\mathbf{N} - \mathbf{c}_N^\mathsf{T})$, given by $d_j = \boldsymbol{\pi}^\mathsf{T}\mathbf{a}_j - c_j$, is zero.

This means that the nonbasic variable may increase from zero without affecting the value of the objective function P. It is of no consequence if it occurs during an intermediate (nonoptimal solution) since we choose the nonbasic variable with the most negative d_j to enter the basis.

However, if it occurs at the optimal solution, then we have many optimal solutions (since P does not change).

Consider the tableau

$$\begin{aligned} P &= \mathbf{c}_B^\mathsf{T}\boldsymbol{\beta} - 0(\mathbf{x}_N)_q \\ \mathbf{x}_B &= \boldsymbol{\beta} - \boldsymbol{\alpha}_q(\mathbf{x}_N)_q \end{aligned} \tag{4-83}$$

$(\mathbf{x}_N)_q$ can increase to a value $\min\limits_{i|\alpha_{iq}>0} (\beta_i/\alpha_{iq}) = \beta_p/\alpha_{pq}$, say, at which time $(\mathbf{x}_N)_q$ enters the basis and $(\mathbf{x}_B)_p$ becomes nonbasic at zero. Thus, any value of $(\mathbf{x}_N)_q$ in the range $0 \leqslant (\mathbf{x}_N)_q \leqslant \beta_p/\alpha_{pq}$ is optimal, as well the two extreme points, when $(\mathbf{x}_N)_q = 0$ and $(\mathbf{x}_B)_p = 0$ respectively.

Note that we obtained a dual degenerate solution and the end of the price-coefficient range in Sec. 4-4-1, and a primal degenerate solution at the end of the right-hand side term range in Sec. 4-4-2.

4-6 EXAMPLE: AN ORE-PROCESSING PLANT

As an example of the wealth of information obtained from postoptimality analysis, we shall consider a simplified industrial problem (we shall also consider the same problem in Sec. 10-3). An ore-processing plant produces two grades of refined product which is sold to the metal industry for further treatment. The operation of the plant is depicted in Fig. 4-1.

There are two types of ore available for processing: type A ore is available up to 100 M tonne/day at a delivered cost of 3.25 $/tonne, while type B ore of higher quality is available up to 30 M tonne/day at a delivered cost of 3.40 $/tonne. The total capacity for basic processing is 100 M tonne/day feedstock at a processing cost of 0.35 $/tonne of feedstock. Basic processing of type A ore produces 15 percent of product I and 85 percent product II, while type B produces 25 percent of product I and 75 percent product II. Product I is more valuable, and a process unit called a converter is able to transform product II into 50 percent product I and 50 percent of a product which can be sold as product II but cannot be reprocessed through the converter. The capacity of the converter is 50 M tonne/day of feedstock, with a conversion cost of 0.25 $/tonne of feedstock. Refining of Product I from basic processing costs 0.10 $/tonne of feedstock.

The market forecast for sales of the products is as follows: product II can be sold in unlimited quantities at 3.80 $/tonne. Product I is sold at 5.50 $/tonne; an existing contract requires that at least 40 M tonne/day of product I is produced, and up to 45 M tonne/day can be sold at this price. Product I inventory can increase at the rate of 4 M tonne/day, and this inventory is valued at 5.20 $/tonne. Excess product I can be sold in unlimited quantities at a distress price of 5.00 $/tonne.

Both products can be purchased if necessary: the delivered cost of product I is 5.75 $/tonne, and the delivered cost of product II is 4.00 $/tonne.

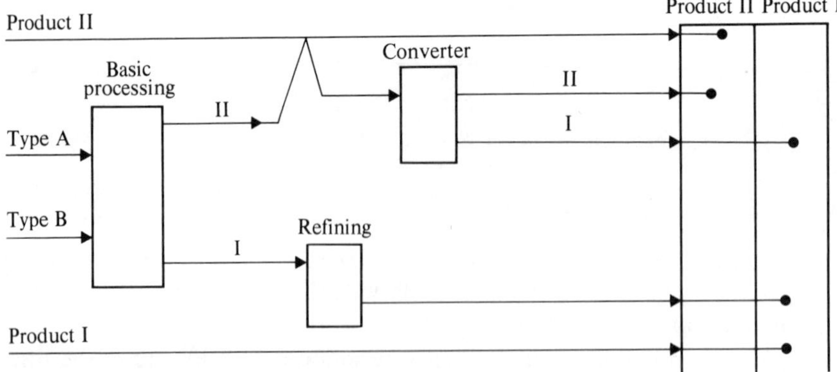

Figure 4-1 Ore-processing plant.

Let us denote the variables as follows (in units of M tonne/day):

x_1 = type A ore processed
x_2 = type B ore processed
x_3 = product I bought
x_4 = product II bought
x_5 = product II sent to converter
x_6 = product I stored as inventory
x_7 = product I sold at distress price
x_8 = slack variable on type B ore availability ($\leqslant 30$)
x_9 = slack variable on upper bound of product I normal sales ($\leqslant 45$)
x_{10} = slack variable on lower bound of product I normal sales ($\geqslant 40$)
x_{11} = slack variable on upper bound on inventory of product I ($\leqslant 4$)
x_{12} = slack variable on upper bound of basic processing capacity ($\leqslant 100$)
x_{13} = slack variable on upper bound of converter capacity ($\leqslant 50$)
x_{14} = excess product II sold directly (i.e., not sent through converter)

After proceeding through the simplex iterations, the final optimal tableau takes the form

Maximize

$$P = 74.3 - 0.44x_{12} - 0.01x_8 - 0.2x_4 - 0.6x_{13} - 0.3x_6 - 0.5x_7 - 0.25x_3$$

subject to

$$\begin{aligned}
x_9 &= 2 + 0.15x_{12} + 0.1x_8 + 0.5x_{13} + 1x_6 + 1x_7 - 1x_3 \\
x_{10} &= 3 - 0.15x_{12} - 0.1x_8 - 0.5x_{13} - 1x_6 - 1x_7 + 1x_3 \\
x_2 &= 30 - 1x_8 \\
x_{11} &= 4 - 1x_6 \\
x_1 &= 70 - 1x_{12} + 1x_8 \\
x_5 &= 50 - 1x_{13} \\
x_{14} &= 32 - 0.85x_{12} + 0.1x_8 + 1x_4
\end{aligned}$$

The operating profit is 74.3 M \$/day, and the corresponding operating plan is given by the values of the variables in the basis. We shall investigate what further information can be gained from a postoptimality analysis.

4-6-1 Shadow Prices

The shadow price on the basic processing capacity constraint is 0.44 \$/tonne (the reduced cost of x_{12} is 0.44). This remains valid over the range $100 + \delta$, where

$$\max\left\{\frac{3}{-0.15}, \frac{70}{-1}, \frac{32}{-0.85}\right\} \leqslant \delta \leqslant \min\left\{\frac{2}{+0.15}\right\}$$

i.e., $-20 \leqslant \delta \leqslant 13.33$.

Thus the operating profit will increase by 0.44 \$/tonne for each tonne increase in basic processing capacity up to 113.33 M tonne/day.

The shadow price on converter capacity is 0.6 \$/tonne (the reduced cost of x_{13} is 0.6). This remains valid over the range $50 + \delta$, where

$$\max\left\{\frac{3}{-0.5}, \frac{50}{-1}\right\} \leqslant \delta \leqslant \min\left\{\frac{2}{0.5}\right\}$$

i.e., $-6 \leqslant \delta \leqslant 4$.

4-6-2 Price Margin

The price margin on type B ore is 0.01 \$/tonne, valid within the range $30 + \delta$, where

$$\max\left\{\frac{3}{-0.1}, \frac{30}{-1}\right\} \leqslant \delta \leqslant \min\left\{\frac{2}{0.1}, \frac{70}{1}, \frac{32}{0.1}\right\}$$

i.e., $-30 \leqslant \delta \leqslant 20$. When $\delta = -30$, then $x_2 = 0$, i.e., no type B ore is purchased. When $\delta = +20$, then $x_2 = 50$, i.e., up to 50 M tonne/day can be purchased, then a change of basis is required.

We conclude that we can make a net profit of 0.01 \$/tonne for each tonne of type B ore which can be purchased in excess of 30 M tonne/day up to a limit of 50 M tonne/day, where the margin changes (because of a change in basis). Also, for each tonne of type B ore less than 30 M tonne/day purchased, we lose 0.01 \$/tonne, all the way down to no purchases. Recall from the discussion in Sec. 4-3 that we could alternatively take the viewpoint that we could negotiate further purchases of type B ore at a price of up to $3.40 + 0.01 = 3.41$ \$/tonne, up to a limit of a further 20 M tonne/day.

4-6-3 Changes in Cost Coefficients

(a) Nonbasic variables

x_4: Reduced cost = 0.2.

If product II can be purchased for $4.00 - 0.2 = 3.80$ \$/tonne or less, it is profitable to do so, in which case we can purchase unlimited quantities of it.

x_6: Reduced cost = 0.3.

If the value of product I stored as inventory is increased to $5.20 + 0.3 = 5.50$ \$/tonne or more, it is profitable to store inventory, and up to $\min\{3/1, 4/1\} = 3$ M tonne/day may be added to inventory (before a change of basis occurs).

x_7: Reduced cost = 0.5.

If the distress price of product I is increased to $5.00 + 0.5 = 5.50$ \$/tonne or more, it is profitable to sell to this market, and 3 M tonne/day can be sold before a change of basis occurs.

x_3: Reduced cost = 0.25.

If product I can be purchased for 5.75 − 0.25 = 5.50 \$/tonne or less it is profitable to do so, in which case up to 2 M tonne/day can be purchased before a change of basis occurs.

(b) Basic variables

x_2: Cost coefficient $c_2 = -3.40$.

The cost coefficient can vary in the range $c_2 + \delta$, where

$$\frac{0.01}{-1} \leqslant \delta \leqslant \infty$$

When the price of type B ore becomes 3.41 \$/tonne or more ($c_2 = -3.40 - 0.01$), it becomes profitable to increase x_8, i.e., reduce the amount of type B ore purchased, and the range x_8 can take is given by

$$\max\left\{\frac{2}{-0.1}, \frac{70}{-1}, \frac{32}{-0.1}\right\} \leqslant x_8 \leqslant \min\left\{\frac{3}{0.1}, \frac{30}{1}\right\}$$

We are normally interested only in the positive limits. In this case x_8 can vary up to 30 M tonne/day before a change of basis is required (x_2 becomes zero). However, examination of the above results shows that they are analogous to the *price margin* calculation performed previously.

This suggests a way of calculating price margins for variables which are not at their upper bound, as we shall see for the case of type A ore (x_1).

x_1: Cost coefficient $c_1 = -3.25$.

The cost coefficient can vary in the range $c_1 + \delta$, where

$$-0.44 \leqslant \delta \leqslant 0.01$$

If the cost of type A ore reduces to 3.24 \$/tonne or less ($c_1 = -3.25 + 0.01$), it becomes profitable to increase x_8, i.e., substitute type A ore for type B ore, up to a limit of $x_8 = 30$, corresponding to $x_1 = 100$, $x_2 = 0$. Thus the *price margin* of type A ore in the range 70–100 M tonne/day is −0.01 \$/tonne.

If the cost of type A ore increases to 3.69 \$/tonne or more ($c_1 = -3.25 - 0.44$), it becomes profitable to increase x_{12}, i.e., reduce the throughput of basic processing by reducing the amount of type A or purchased, up to a limit of $x_{12} = 3/0.15 = 20$, corresponding to $x_1 = 50$. Thus the *price margin* of type A ore in the range 50–70 M tonne/day is 0.44 \$/tonne.

Note that the break-point for price margins of a basic variable is the level at which the variable is in the optimum solution (in this case 70). The price margins are interpreted slightly differently as being the price changes required to produce a change in operating plan: if type A ore can be purchased for 0.01 \$/tonne less, then it is equally profitable to substitute it for type B ore, and if the cost increases by 0.44 \$/tonne, then it becomes equally profitable to decrease purchases, by up to 20 M tonne/day.

4-6-4 Changes in Right-hand Side Terms

(a) Basic slack variable The amount of change can be calculated directly:

x_9 and x_{10} indicate that the level of product I normal sales could be specified as 43 (upper bound decreased by 2 and lower bound increased by 3).

x_{11} indicates that the upper limit of product I to inventory could be decreased by 4.

(b) Nonbasic slack variable The ranges of change have all been discussed with shadow prices, price margins, and changes in the cost coefficients. However, we may wish to examine separately the effect of changes in availability quite regardless of changes in price. We will thus briefly summarize the effects of right-hand side changes alone, and examine the range within which the current optimum remains optimal and feasible.

x_{12}: Basic processing capacity, can vary in the range $100 + \delta$, where

$$\max\left\{\frac{3}{-0.15}, \frac{70}{-1}, \frac{32}{-0.85}\right\} \leqslant \delta \leqslant \min\left\{\frac{2}{0.15}\right\}$$

i.e., $-20 \leqslant \delta \leqslant 13.33$

x_8: Type B ore availability, can vary in the range $30 + \delta$, where

$$\max\left\{\frac{3}{-0.1}, \frac{30}{-1}\right\} \leqslant \delta \leqslant \min\left\{\frac{2}{0.1}, \frac{70}{1}, \frac{32}{0.1}\right\}$$

i.e., $-30 \leqslant \delta \leqslant 20$

x_{13}: Converter capacity, can vary in the range $50 + \delta$, where

$$\max\left\{\frac{3}{-0.5}, \frac{50}{-1}\right\} \leqslant \delta \leqslant \min\left\{\frac{2}{0.5}\right\}$$

i.e., $-6 \leqslant \delta \leqslant 4$.

Recall from Sec. 4-4-2 that optimality is preserved in the sense that the basis does not change, even though the values of x_B (and P) change, but remain feasible.

4-7 THE DUAL-SIMPLEX METHOD

This procedure essentially applies the simplex method to the dual problem while working with the primal tableau. Suppose we have the standard primal tableau, expressed in the form

$$\begin{aligned} \text{Maximize} \quad & P = \mathbf{c}_B^{\top}\boldsymbol{\beta} - d_1(\mathbf{x}_N)_1 - \cdots - d_q(\mathbf{x}_N)_q - \cdots - d_{n-m}(\mathbf{x}_N)_{n-m} \\ \text{subject to} \quad & \mathbf{x}_B = \boldsymbol{\beta} - \boldsymbol{\alpha}_1(\mathbf{x}_N)_1 - \cdots - \boldsymbol{\alpha}_q(\mathbf{x}_N)_q - \cdots - \boldsymbol{\alpha}_{n-m}(\mathbf{x}_N)_{n-m} \end{aligned} \tag{4-84}$$

Suppose further that the tableau is dual feasible (i.e., optimal) but not necessarily primal feasible.

The steps of the dual-simplex method are as follows:

1. Choose the pivot row p.

Select
$$\beta_p = \min_{i \mid \beta_i < 0} \{\beta_i\} \tag{4-85}$$

(*a*) If none exist, STOP; we have an optimal and feasible solution

(*b*) Otherwise let the minimum correspond to $i = p$, say. $(\mathbf{x}_B)_p$ is the leaving basic variable.

2. To choose the entering nonbasic variable, we first of all note that the reduced cost of $(\mathbf{x}_B)_p$ in the new tableau if $(\mathbf{x}_N)_j$ replaces $(\mathbf{x}_B)_p$ in the basis is given by

$$\bar{d}_j = \frac{-d_j}{(\boldsymbol{\alpha}_j)_p} \tag{4-86}$$

(The reader may wish to confirm Eq. (4-86) by using Eqs. (2-38) and (2-39) on the cost row term (d_j) considered as part of $\boldsymbol{\alpha}_j$, cf. Eq. (4-84), with $\mathbf{Be}_p$ replacing $\boldsymbol{\alpha}_j$ in the nonbasic portion of the matrix.)

The ratio test is

$$\theta = \min_{j \mid (\boldsymbol{\alpha}_j)_p < 0} \left\{ \frac{-d_j}{(\boldsymbol{\alpha}_j)_p} \right\} \tag{4-87}$$

for $j = q$, say.

If any higher ratio was used, say for $j = r$, and $(\mathbf{x}_N)_r$ entered the basis, then again using Eqs. (2-38) and (2-39) we have

$$\bar{d}_q = d_q - \frac{(\boldsymbol{\alpha}_q)_p}{(\boldsymbol{\alpha}_r)_p} d_r \tag{4-88}$$

$$< 0, \quad \text{if } \frac{-d_r}{(\boldsymbol{\alpha}_r)_p} > \frac{-d_q}{(\boldsymbol{\alpha}_q)_p}$$

and optimality would be lost.

From Eq. (2-39) it is evident that, when $(\mathbf{x}_N)_q$ enters the basis,

$$\bar{d}_j = d_j + \theta(\boldsymbol{\alpha}_j)_p, j \neq q \tag{4-89}$$

3. If there is no element $(\boldsymbol{\alpha}_j)_p < 0$ for any of the nonbasic variables $(\mathbf{x}_N)_j$, then no nonnegative value of $(\mathbf{x}_N)_j$ can cause $(\mathbf{x}_B)_p$ to be feasible: STOP, the problem is infeasible.

Otherwise, we perform a pivot operation on the primal, interchanging $(\mathbf{x}_N)_q$ with $(\mathbf{x}_B)_p$ in the basis, and repeat from Step 1.

We have two terminating criteria as in the (primal) simplex method, however, note that the second criterion (infeasibility) is the dual of that of the primal method (unboundedness).

The motivation for using the dual-simplex method is twofold:

1. Adding extra rows to the primal problem. Using the dual-simplex method, this is achieved with essentially the same ease as adding an extra column to the primal and pricing out the new column with the current basis.
2. It is also used in parametric programming of the right-hand side vector, which is discussed in Sec. 5-3.

CHAPTER

FIVE

EXTENSIONS TO THE SIMPLEX METHOD

5-1 UPPER BOUNDING

If we replace $\mathbf{x} \geqslant \mathbf{0}$ in Eq. (2-3) by $\mathbf{0} \leqslant \mathbf{x} \leqslant \mathbf{u}$, we refer to this as *upper bounding*.

This extension to the simplex method was developed very early and it could be argued that it should have been included in our discussion of the revised simplex method. However, it is placed here as it provides a good framework for discussing generalized upper bounding, which is a relatively recent development.

The main concept behind upper bounding is that any nonbasic variable x_j may either be zero *or* at its upper bound u_j. This eliminates the need to represent the upper bound on a variable as a separate constraint, thus saving a row for each upper bound. This reduces m, the size of the basis matrix, with obvious resultant savings in both running time and storage requirements. The price paid for such savings is that the steps of the simplex algorithm become slightly more complicated.

Suppose the problem is of the form

Maximize $$P = \mathbf{c}^{\top}\mathbf{x} \tag{5-1}$$

subject to $$\mathbf{A}\mathbf{x} = \mathbf{b} \tag{5-2}$$

$$\mathbf{0} \leqslant \mathbf{x} \leqslant \mathbf{u} \tag{5-3}$$

(Nonzero lower bounds can be accommodated effectively; however, it is more convenient for our discussion to assume they are all zero.) Now this problem could be solved by the standard form of the revised simplex method by expressing the constraints in the form

$$\mathbf{Ax} = \mathbf{b} \tag{5-4}$$

$$\mathbf{x} + \mathbf{x}_s = \mathbf{u} \tag{5-5}$$

$$\mathbf{x} \geqslant \mathbf{0},\ \mathbf{x}_s \geqslant \mathbf{0} \tag{5-6}$$

i.e.,

$$\left[\begin{array}{c|c} \mathbf{A} & \mathbf{0} \\ \hline \mathbf{I} & \mathbf{I} \end{array}\right] \left[\begin{array}{c} \mathbf{x} \\ \hline \mathbf{x}_s \end{array}\right] = \left[\begin{array}{c} \mathbf{b} \\ \hline \mathbf{u} \end{array}\right] \tag{5-7}$$

$$\mathbf{x} \geqslant \mathbf{0},\ \mathbf{x}_s \geqslant \mathbf{0}$$

where $\mathbf{x}_s$ is the vector of slack variables associated with the upper bounds. However, we have increased the number of rows (and number of columns) by n.

We can eliminate the need to do this by allowing the nonbasic variables to be at *either* end of their range, as given by Eq. (5-3). The only consequence of this is that:

1. If $x_j = 0$, we consider the possibility of increasing it.
2. If $x_j = u_j$, we consider the possibility of decreasing it.

Remembering that $P = \mathbf{c}_B^\top \mathbf{B}^{-1}\mathbf{b} - d_1(\mathbf{x}_N)_1 - \cdots - d_{n-m}(\mathbf{x}_N)_{n-m}$ we see that if $x_j = 0$ and the corresponding reduced cost is negative, P will increase on increasing x_j above zero, while if $x_j = u_j$ and the corresponding reduced cost is positive, P will increase on decreasing x_j below u_j.

Thus Step 3 of the revised simplex method is modified to:

1. Choose the most negative d_j among the nonbasic variables at zero.
2. Choose the most positive d_j among the nonbasic variables at their upper bound.
3. If none exist (within the error tolerance TOLDJ), STOP; otherwise, choose q corresponding to the reduced cost of largest magnitude in (1) and (2).

Having chosen a nonbasic variable $(\mathbf{x}_N)_q$ (to increase if zero, to decrease if at its upper bound, say u_q), we now produce

$$\boldsymbol{\alpha}_q = \mathbf{B}^{-1}\mathbf{a}_q \tag{5-8}$$

as in Step 4 of the simplex method.

Consider the tableau equation

$$\mathbf{x}_B = \boldsymbol{\beta} - \boldsymbol{\alpha}_1(\mathbf{x}_N)_1 - \cdots - \boldsymbol{\alpha}_q(\mathbf{x}_N)_q - \cdots - \boldsymbol{\alpha}_{n-m}(\mathbf{x}_N)_{n-m} \tag{5-9}$$

Since the nonbasic variables may be at their upper bound, $\boldsymbol{\beta}$ does *not* represent the current value of $\mathbf{x}_B$. Instead, we have

$$\mathbf{x}_B = \boldsymbol{\beta} - \sum_{j \in S} \boldsymbol{\alpha}_j(\mathbf{x}_N)_j \tag{5-10}$$

where S is the set of nonbasic variables at their upper bound; we must consider their contribution to the current value of $\mathbf{x}_B$.

To distinguish between current and new values, let $(\mathbf{x}_B)_i$ denote the current value of the ith basic variable, and x_i denote its new value as $(\mathbf{x}_N)_q$ changes.

As $(\mathbf{x}_N)_q$ increases (decreases), each element $\mathbf{x}_i$ will change:

1. If α_{iq} is positive, x_i will decrease (increase).
2. If α_{iq} is negative, x_i will increase (decrease).

Thus in modifying Step 5 of the simplex method, we must consider three possible outcomes:

(*a*) A basic variable may reach its lower bound (zero) first.
(*b*) A basic variable may reach its upper bound first.
(*c*) The nonbasic variable may reach the other end of its range before (*a*) or (*b*) occur.

Suppose $(\mathbf{x}_N)_q$ is increased from zero:
For (*a*) we have

$$(\mathbf{x}_B)_i - \alpha_{iq}(\mathbf{x}_N)_q \geqslant 0 \tag{5-11}$$

thus

$$(\mathbf{x}_N)_q \leqslant \min_{i|\alpha_{iq}>0} \left\{ \frac{(\mathbf{x}_B)_i}{\alpha_{iq}} \right\} \tag{5-12}$$

(cf. Eq. (2-29))
For (*b*) we have

$$(\mathbf{x}_B)_i - \alpha_{iq}(\mathbf{x}_N)_q \leqslant u_i \tag{5-13}$$

thus

$$(\mathbf{x}_N)_q \leqslant \min_{i|\alpha_{iq}<0} \left\{ \frac{u_i - (\mathbf{x}_B)_i}{-\alpha_{iq}} \right\} \tag{5-14}$$

For (*c*) we have

$$(\mathbf{x}_N)_q \leqslant u_q \tag{5-15}$$

thus the value that $(\mathbf{x}_N)_q$ can take is the smallest of these,

$$(\mathbf{x}_N)_q = \min \left\{ \min_{i|\alpha_{iq}>0} \left\{ \frac{(\mathbf{x}_B)_i}{\alpha_{iq}} \right\}, \min_{i|\alpha_{iq}<0} \left\{ \frac{u_i - (\mathbf{x}_B)_i}{-\alpha_{iq}} \right\}, u_q \right\} \tag{5-16}$$

Now suppose $(\mathbf{x}_N)_q$ is decreased from u_q; we have the current values of the basic variables given by

$$(\mathbf{x}_B)_i = \beta_i - \sum_{\substack{j \in S \\ j \neq q}} \alpha_{ij}(\mathbf{x}_N)_j - \alpha_{iq} u_q \tag{5-17}$$

Thus for any new value of $(\mathbf{x}_N)_q$ we have

$$x_i = (\mathbf{x}_B)_i + \alpha_{iq}(u_q - (\mathbf{x}_N)_q) \tag{5-18}$$

Defining the decrease in the nonbasic variable by $\delta = u_q - (\mathbf{x}_N)_q$:
For (a) we have

$$(\mathbf{x}_B)_i + \alpha_{iq}(+\delta) \geqslant 0 \tag{5-19}$$

thus
$$\delta \leqslant \min_{i|\alpha_{iq}<0} \left\{ \frac{(\mathbf{x}_B)_i}{-\alpha_{iq}} \right\} \tag{5-20}$$

For (b) we have

$$(\mathbf{x}_B)_i + \alpha_{iq}(+\delta) \leqslant u_i \tag{5-21}$$

thus
$$\delta \leqslant \min_{i|\alpha_{iq}>0} \left\{ \frac{u_i - (\mathbf{x}_B)_i}{\alpha_{iq}} \right\} \tag{5-22}$$

For (c) we have $\delta \leqslant u_q$
Thus the value that δ can take is the smallest of these,

$$\delta = \min \left\{ \min_{i|\alpha_{iq}<0} \left\{ \frac{(\mathbf{x}_B)_i}{-\alpha_{iq}} \right\}, \min_{i|\alpha_{iq}>0} \left\{ \frac{u_i - (\mathbf{x}_B)_i}{\alpha_{iq}} \right\}, u_q \right\} \tag{5-23}$$

and
$$(\mathbf{x}_N)_q = u_q - \delta \tag{5-24}$$

If (c) occurs, no change of basis is required; we merely record the fact that $(\mathbf{x}_N)_q$ has gone to the other end of its range and adjust the value of $\mathbf{x}_B$ to $\bar{\mathbf{x}}_B$, where

$$\bar{\mathbf{x}}_B = \mathbf{x}_B - \boldsymbol{\alpha}_q(\pm u_q) \tag{5-25}$$

where the appropriate sign is: $\begin{cases} + \text{ if } (\mathbf{x}_N)_q \text{ increased from zero} \\ - \text{ if } (\mathbf{x}_N)_q \text{ decreased from } u_q \end{cases}$

If (a) or (b) occur, for say $i = p$, the appropriate pivot operation corresponding to the change of basis (Step 6) is made, and $\mathbf{x}_B$ adjusted according to

$$\bar{\mathbf{x}}_B = \begin{cases} (\mathbf{x}_B)_i - \alpha_{iq}(\mathbf{x}_N)_q & i \neq p;\ (\mathbf{x}_N)_q \text{ increased from zero} \\ (\mathbf{x}_B)_i + \alpha_{iq}(+\delta) & i \neq p;\ (\mathbf{x}_N)_q \text{ decreased from } u_q \\ (\mathbf{x}_N)_q & i = p \end{cases} \tag{5-26}$$

A record of which nonbasic variables are at their upper bound must be maintained.

Lower bounds $\mathbf{x} \geqslant \mathbf{l}$ can be removed a priori by the transformation $\mathbf{x} = \bar{\mathbf{x}} + \mathbf{l}$. Thus the problem

Maximize	$\mathbf{c}^\mathsf{T}\mathbf{x}$
subject to	$\mathbf{A}\mathbf{x} = \mathbf{b}$
	$\mathbf{l} \leqslant \mathbf{x} \leqslant \mathbf{u}$

Changes to

Maximize $\qquad \mathbf{c}^{\mathsf{T}}\bar{\mathbf{x}} + \mathbf{c}^{\mathsf{T}}\mathbf{l}$ (constant)

subject to
$$\mathbf{A}\bar{\mathbf{x}} = \mathbf{b} - \mathbf{A}\mathbf{l}$$
$$\mathbf{0} \leqslant \bar{\mathbf{x}} \leqslant \mathbf{u} - \mathbf{l}$$

However, such a transformation can be tiresome to expedite, and most commercial codes offer the convenience of allowing variables to be nonbasic at their (nonzero) lower bounds, which are accounted for in an analogous manner to upper bounds.

5-1-1 Calculation of Price Margins

Recall that in Sec. 4-3 the price margin of a variable was calculated as the shadow price of the upper-bound constraint. Using the upper-bounding technique discussed here, the constraint does not appear explicitly; however, the same result appears as the reduced cost of the variable itself, which will be nonbasic at its upper bound if the constraint is limiting.

This reduced cost will be negative at the optimum, and so must be multiplied by -1 to get the price margin.

5-2 GENERALIZED UPPER BOUNDING (GUB)

Generalized upper bounding is a generalization of the upper bounding principle to include upper bounds on *groups* of variables rather than single variables. The structure of problems on which generalized upper bounding can be used is of the following form:

Maximize $\qquad P = \mathbf{c}_0^{\mathsf{T}}\mathbf{x}_0 + \mathbf{c}_1^{\mathsf{T}}\mathbf{x}_1 + \cdots + \mathbf{c}_k^{\mathsf{T}}\mathbf{x}_k + \cdots + c_t^{\mathsf{T}}\mathbf{x}_t = b_0 \qquad (5\text{-}27)$

subject to: $\qquad \mathbf{A}_0\mathbf{x}_0 + \mathbf{A}_1\mathbf{x}_1 + \cdots + \mathbf{A}_k\mathbf{x}_k + \cdots + \mathbf{A}_t\mathbf{x}_t = \mathbf{b} \qquad (5\text{-}28)$

$$\begin{aligned} \mathbf{1}^{\mathsf{T}}\mathbf{x}_1 \qquad\qquad\qquad\qquad &= b_{m+1} \\ \ddots \qquad\qquad &\ \ \vdots \\ \mathbf{1}^{\mathsf{T}}\mathbf{x}_k \qquad\quad &= b_{m+k} \qquad (5\text{-}29) \\ \ddots \qquad &\ \ \vdots \\ \mathbf{1}^{\mathsf{T}}\mathbf{x}_t &= b_{m+t} \end{aligned}$$

(all variables nonnegative)

The last t tows are called *convexity rows* or *GUB rows*, and the variables occurring in such rows are referred to as *GUB sets*. $\mathbf{A}_k = [a_{ijk}]$ and $\mathbf{x}_k = [x_{jk}]$. The extra subscript k refers to the kth GUB set; some variables are not associated with GUB sets, and these may be shown separately as in the far left-hand side of Eqs. (5-27) and (5-28), or simply referred to as corresponding to $k = 0$. The GUB rows have the structure

$$\begin{bmatrix} 1 & 1 & 1 & & & & & & & & & & & \\ & & & 1 & 1 & & & & & & & & & \\ & & & & & \ddots & & & & & & & & \\ & & & & & & 1 & 1 & 1 & & & & & \\ & & & & & & & & & \ddots & & & & \\ & & & & & & & & & & 1 & 1 & 1 & 1 & 1 \end{bmatrix} \begin{bmatrix} \mathbf{x}_1 \\ \mathbf{x}_2 \\ \vdots \\ \mathbf{x}_k \\ \vdots \\ \mathbf{x}_t \end{bmatrix} = \begin{bmatrix} b_{m+1} \\ b_{m+2} \\ \vdots \\ b_{m+k} \\ \vdots \\ b_{m+t} \end{bmatrix} \tag{5-30}$$

Such structures appear commonly in practice, particularly in multi-divisional or multi-time period production and distribution problems where there are availability, capacity, or demand constraints.

For example, consider an organization with t divisions, the kth having b_{m+k} man-hours available for $J(k)$ projects. If x_{jk} is the number of man-hours devoted to project j, then we have

$$x_{1k} + x_{2k} + \cdots + x_{jk} + \cdots + x_{J(k)k} = b_{m+k} \tag{5-31}$$

$J(k)$ is the number of variables in the kth GUB set.

Generalized upper bounding, developed by Dantzig and Van Slyke (1967), exploits the structure of these constraints. The important attribute is that no variable occurs in more than one GUB set. (Nonunit entries in the GUB row can be made so by suitable scaling of the variable, e.g., if the coefficient of x_{jk} is 2.5, define $\bar{x}_{jk} = 2.5x_{jk}$.) The idea behind the method is that *at least one* variable from each GUB set must be basic, otherwise the basis matrix would have a zero row and be singular.

One such variable is called the *key variable*, and is denoted by $x_{\bar{j}k}$ for each $k = 1, \ldots, t$. The GUB rows are then used to eliminate the key variables from the ordinary rows.

We thus substitute

$$x_{\bar{j}k} = b_{m+k} - \sum_{\substack{j=1 \\ j \neq \bar{j}}}^{J(k)} x_{jk} \qquad k = 1, \ldots, t \tag{5-32}$$

in each row $i = 0, 1, \ldots, m$.

The original problem, Eqs. (5-27)–(5-29), then becomes

Maximize

$$P = \sum_{j=1}^{J(0)} c_j x_j + \sum_{k=1}^{t} \left[\sum_{\substack{j=1 \\ j \neq \bar{j}}}^{J(k)} c_{jk} x_{jk} + c_{\bar{j}k} \left(b_{m+k} - \sum_{\substack{j=1 \\ j \neq \bar{j}}}^{J(k)} x_{jk} \right) \right] = b_0 \tag{5-33}$$

subject to

$$\sum_{j=1}^{J(0)} a_{ij}x_j + \sum_{k=1}^{t}\left[\sum_{\substack{j=1\\ j\neq \bar{j}}}^{J(k)} a_{ijk}x_{jk} + a_{i\bar{j}k}\Big(b_{m+k} - \sum_{\substack{j=1\\ j\neq \bar{j}}}^{J(k)} x_{jk}\Big)\right] = b_i \qquad i = 1, \ldots, m \tag{5-34}$$

The ordinary rows of the problem thus become

Maximize $$P = \mathbf{c}_0^{\mathsf{T}}\mathbf{x}_0 + \mathbf{c}_1^{*\mathsf{T}}\mathbf{x}_1 + \cdots + \mathbf{c}_k^{*\mathsf{T}}\mathbf{x}_k + \cdots + \mathbf{c}_t^{*\mathsf{T}}\mathbf{x}_t = b_0^* \tag{5-35}$$

subject to $$\mathbf{A}_0\mathbf{x}_0 + \mathbf{A}_1^*\mathbf{x}_1 + \cdots + \mathbf{A}_k^*\mathbf{x}_k + \cdots + \mathbf{A}_t^*\mathbf{x}_t = \mathbf{b}^* \tag{5-36}$$

where $$\mathbf{c}_k^* = [c_{jk}^*] = [c_{jk} - c_{\bar{j}k}] \tag{5-37}$$

$$\mathbf{A}_k^* = [a_{ijk}^*] = [a_{ijk} - a_{i\bar{j}k}] \tag{5-38}$$

$$\mathbf{b}^* = [b_i^*] = \left[b_i - \sum_{k=1}^{t} a_{i\bar{j}k}b_{m+k}\right] \tag{5-39}$$

and $$b_0^* = b_0 - \sum_{k=1}^{t} c_{i\bar{j}k}b_{m+k} \tag{5-40}$$

We have now effectively eliminated the GUB rows from the problem. We could, in theory, work with this shortened problem and use Eq. (5-31) to solve for the values of the key variables. However, the key variables themselves are not allowed to go negative, so they must be included for consideration when performing the ratio test to find the leaving basic variable. A discussion of the generalized upper bound algorithm follows (Beale, 1970).

Let R_k be the set of variables in the kth GUB row which are nonbasic, and S_k be the set of variables in the kth GUB row which are basic (excluding the key). Suppose the current tableau for the *shortened* problem is as follows

Maximize $$P = \beta_0 - \sum_{k=0}^{t}\sum_{j\in R_k} d_{jk}x_{jk} \tag{5-41}$$

subject to $$\mathbf{x}_B = \boldsymbol{\beta} - \sum_{k=0}^{t}\sum_{j\in R_k} \boldsymbol{\alpha}_{jk}x_{jk} \tag{5-42}$$

(i.e., $x_i = \beta_i - \sum_{k=0}^{t}\sum_{j\in R_k} \alpha_{ijk}x_{jk} \qquad i = 1, \ldots, m$)

d_{jk} is the reduced cost of the nonbasic variable x_{jk}; β_0 and β_i, $i = 1, \ldots, m$ are the current values of P and x_i, $i = 1, \ldots, m$ respectively. Consider a particular GUB row, $k = h$, say.

We have from Eq. (5-32)

$$x_{\bar{j}h} = b_{m+h} - \sum_{\substack{j=1\\ j\neq \bar{j}}}^{J(h)} x_{jh} \tag{5-43}$$

Now, S_h is the set of variables in the hth GUB row which are basic. These are in the tableau in the form

$$x_i = \beta_i - \sum_{k=0}^{t} \sum_{j \in R_k} \alpha_{ijk} x_{jk} \qquad i \in S_h \tag{5-44}$$

Let us put this expression in Eq. (5-43).

The original hth GUB row, Eq. (5-43), becomes

$$\begin{aligned} x_{jh} &= b_{m+h} - \sum_{i \in S_h} \left(\beta_i - \sum_{k=0}^{t} \sum_{j \in R_k} \alpha_{ijk} x_{jk} \right) - \sum_{j \in R_h} x_{jh} \\ &= b_{m+h} - \sum_{i \in S_h} \left(\beta_i - \sum_{k=0}^{t} \sum_{j \in R_k} \alpha_{ijk} x_{jk} \right) - \sum_{k=0}^{t} \sum_{j \in R_k} \delta_{kh} x_{jk} \end{aligned} \tag{5-45}$$

where $$\delta_{kh} = \begin{cases} 1 & k = h \\ 0 & k \neq h \end{cases}$$

We have $$x_{jh} = \left(b_{m+h} - \sum_{i \in S_h} \beta_i \right) - \sum_{k=0}^{t} \sum_{j \in R_k} \left(\delta_{kh} - \sum_{i \in S_h} \alpha_{ijk} \right) x_{jk} \tag{5-46}$$

Thus the "bottom half" of the tableau has equations of the form

$$x_{jh} = \beta_{m+h} - \sum_{k=0}^{t} \sum_{j \in R_k} \alpha_{m+h,\,j,\,k} x_{jk} \tag{5-47}$$

where $$\beta_{m+h} = b_{m+h} - \sum_{i \in S_h} \beta_i \qquad h = 1, \ldots, t \tag{5-48}$$

$$\alpha_{m+h,\,j,\,k} = \delta_{kh} - \sum_{i \in S_h} \alpha_{ijk} \qquad j \in R_k,\ k = 0, \ldots, t,\ h = 1, \ldots, t \tag{5-49}$$

Note that this is of the same form as the shortened problem, i.e., the "top half."

The complete tableau, Eqs. (5-41), (5-42), and (5-47), is used in evaluating the reduced costs of the nonbasic variables, and in performing the ratio test to find the leaving basic variable. However, it should be emphasized that the "bottom half," Eq. (5-47), is only generated as required, and we only need to work with the basis **B** formed from the columns of the shortened problem, Eqs. (5-35) and (5-36).

5-2-1 Steps of the Procedure

Step 1: Produce pricing vector Recall that in the ordinary simplex method we produce $\boldsymbol{\pi}^\top = \mathbf{c}_B^\top \mathbf{B}^{-1}$, and the reduced cost of the jth nonbasic variable is given by

$$\mathbf{c}_B^\top \mathbf{B}^{-1} \mathbf{a}_j - c_j = \mathbf{c}_B^\top \boldsymbol{\alpha}_j - c_j = \sum_{i=1}^{m} c_i \alpha_{ij} - c_j \tag{5-50}$$

where c_i is the ith component of $\mathbf{c}_B$.

In the generalized upper bounding formulation, the reduced cost of the jkth nonbasic variable is given by

$$d_{jk} = \sum_{i=1}^{m} c_i \alpha_{ijk} + \sum_{h=1}^{t} c_{m+h} \alpha_{m+h,\,j,\,k} - c_{jk} \tag{5-51}$$

where c_i is the ith component of $\mathbf{c}_B$, the m-dimensional basis subvector of $\mathbf{c}^\top = (c_j, j = 1, \ldots, J(0);\ c_{jk}^*, j = 1, \ldots, J(k),\ j \neq \bar{j},\ k = 1, \ldots, t)$ corresponding to Eq. (5-35), and c_{m+h} is the cost coefficient of the hth key variable $x_{\bar{j}h}$. Using Eq. (5-49),

$$\begin{aligned} d_{jk} &= \sum_{i=1}^{m} c_i \alpha_{ijk} + \sum_{h=1}^{t} c_{m+h} \left(\delta_{kh} - \sum_{i \in S_h} \alpha_{ijk} \right) - c_{jk} \\ &= \sum_{i=1}^{m} (c_i - \bar{c}_i) \alpha_{ijk} + c_{m+k} - c_{jk} \end{aligned} \tag{5-52}$$

where
$$\bar{c}_i = \begin{cases} 0 \text{ if } i \notin S_h,\ h = 1, \ldots, t \\ c_{m+h} \text{ if } i \in S_h,\ h = 1, \ldots, t \end{cases}$$

Therefore, to produce the pricing vector, $\boldsymbol{\pi}^\top$, the components of the row vector to premultiply into $\mathbf{B}^{-1}$ are $(c_i - \bar{c}_i)$, $i = 1, \ldots, m$.

Step 2: Price out the nonbasic columns

$$d_{jk} = \boldsymbol{\pi}^\top \mathbf{a}_{jk}^* + c_{m+k} - c_{jk} \tag{5-53}$$

Step 3: Select entering nonbasic variable Choose the most negative reduced cost:

1. If none, we have an optimal solution.
2. Otherwise, let the entering nonbasic variable be $jk = (qr)$, say.

Step 4: Produce updated $\boldsymbol{\alpha}_{qr}$ Produce

$$\boldsymbol{\alpha}_{qr} = \mathbf{B}^{-1} \mathbf{a}_{qr}^* \tag{5-54}$$

Step 5: Find leaving basic variable Choose the minimum positive ratio β_i/α_{iqr}, $i = 1, \ldots, m + t$. Note that this includes the top half *and* the bottom half, using Eqs. (5-47)–(5-49) for the latter.

1. If none, then the solution is unbounded.
2. Otherwise, let the leaving basic variable be $i = p$, say.

Step 6: Pivot operation

1. If $p \leq m$, we perform the standard pivot operation, replacing the pth column of $\mathbf{B}$ with $\mathbf{a}_{qr}^*$.
2. If $p > m$, a *key variable* is removed from the basis.

Recall we have for $p > m$

$$x_{\bar{j}h} = \beta_{m+h} - \sum_k \sum_j \alpha_{m+h,j,k} x_{jk} \tag{5-47}$$

where
$$\alpha_{m+h,j,k} = \delta_{kh} - \sum_{i \in S_h} \alpha_{ijk}$$

Two possibilities arise:

(*a*) Suppose the key variable for the rth GUB set must be removed, i.e., $p = m + r$ and *no* other variable in the rth GUB set is basic (i.e., S_r is null). **B** does not change.

Note that

$$x_i = \beta_i - \alpha_{iqr} x_{qr}, \qquad i = 1, \ldots, m.$$

Also, $x_{\bar{j}r} = b_{m+r} - 1x_{qr}$, so $x_{qr} = b_{m+r}$ when $x_{\bar{j}r}$ goes to zero. Therefore

$$x_i = \beta_i - \alpha_{iqr} b_{m+r}, \qquad i = 1, \ldots, m$$

when x_{qr} becomes the new key.

We thus merely insert the new key, and subtract $\alpha_{iqr} b_{m+r}$ from each β_i, $i = 1, \ldots, m$ (and $d_{qr} b_{m+r}$ from β_0).

(*b*) Suppose the key variable for the sth GUB set must be removed, and other variables in this set are basic. (Note that the index s is not necessarily equal to the index r, i.e., the entering nonbasic variable x_{qr} does not necessarily belong to the same GUB set as the leaving key variable, which is in sth GUB set.)

Since other variables in the sth GUB set are basic, *one* of these is chosen to be the new key variable, and the *other* basic variables in the sth GUB set have their corresponding columns in **B** change, since the columns $\mathbf{a}_{js}^*$ will change on changing the key variable.

The procedure then is to perform a number of pivot (column replacement) operations to replace the old columns in **B** with the new columns $\mathbf{a}_{js}^*$ corresponding to the sth GUB set; however, it turns out that these interchanges give rise to particularly simple transformations.

Although the procedure seems complicated, the resultant saving in storage and running time is well worth the effort, especially if the GUB rows constitute a major proportion of the rows. Orchard-Hays (1974) reports that the reduction in total running time approaches a factor of 10 when the proportion of GUB rows is around 80 percent.

5-3 PARAMETRIC PROGRAMMING

In Sec. 4-5 we discussed how the optimal solution of a linear-programming problem changes as either the cost coefficients or the right-hand side elements change. However, the discussion was limited to the range within which the current basis stays optimal, and considered the effect of changing only one term at a time.

Parametric programming is a method of determining how the solution of a problem varies as either the objective row or right-hand side vary.

5-3-1 Parametric Variation of the Cost Row

We may state the problem in the form

$$\begin{array}{lll} \text{Maximize} & P + \theta\hat{P} & \\ \text{in the range} & 0 \leqslant \theta \leqslant \theta_{\max} & \\ \text{where} & P = \mathbf{c}^{\mathsf{T}}\mathbf{x} & \text{the original objective function} \\ & \hat{P} = \hat{\mathbf{c}}^{\mathsf{T}}\mathbf{x} & \text{the parametric terms} \\ \text{subject to} & \mathbf{A}\mathbf{x} = \mathbf{b} & \\ & \mathbf{x} \geqslant \mathbf{0} & \end{array} \tag{5-55}$$

Considering only the original objective function ($\theta = 0$), suppose the optimal tableau is of the form (cf. Eqs. (2-25) and (2-26))

$$\begin{aligned} P &= \beta_0 - d_1(\mathbf{x}_N)_1 - \cdots - d_j(x_N)_j - \cdots \\ \hat{P} &= \hat{\beta}_0 - \hat{d}_1(x_N)_1 - \cdots - \hat{d}_j(x_N)_j - \cdots \\ \mathbf{x}_B &= \boldsymbol{\beta} - \boldsymbol{\alpha}_1(x_N)_1 - \cdots - \boldsymbol{\alpha}_j(x_N)_j - \cdots \end{aligned} \tag{5-56}$$

where $\beta_0 = \mathbf{c}_B^{\mathsf{T}}\mathbf{B}^{-1}\mathbf{b}$ and $\hat{\beta}_0 = \hat{\mathbf{c}}_B^{\mathsf{T}}\mathbf{B}^{-1}\mathbf{b}$.

Here the cost row $\hat{P}$ was calculated using the optimal basis for P (so the terms $\hat{d}_j$ may not necessarily be positive), where $\hat{d}_j = \hat{\mathbf{c}}_B^{\mathsf{T}}\mathbf{B}^{-1}\mathbf{a}_j - \hat{c}_j$.

However, since the tableau is optimal with respect to P we have

$$\begin{aligned} \beta_i &\geqslant 0 \qquad i = 1, \ldots, m \\ d_j &\geqslant 0 \qquad j = 1, \ldots, n - m \end{aligned} \tag{5-57}$$

As θ is increased from zero, the composite objective function becomes

$$(\beta_0 + \theta\hat{\beta}_0) - \sum_{j=1}^{n-m} (d_j + \theta\hat{d}_j)(\mathbf{x}_N)_j \tag{5-58}$$

This solution remains optimal as long as

$$d_j + \theta\hat{d}_j \geqslant 0 \qquad j = 1, \ldots, n - m \tag{5-59}$$

We thus compute

$$\theta_c = \min_{j \mid \hat{d}_j < 0} \left\{ \frac{d_j}{-\hat{d}_j} \right\} \tag{5-60}$$

for $j = q$, say.

This minimum positive ratio gives the largest value of θ for which the solution remains optimal, since for $\theta > \theta_c$, the reduced cost of $(\mathbf{x}_N)_q$ is negative. At $\theta = \theta_c$

we therefore perform a change of basis:

1. $(\mathbf{x}_N)_q$ is made basic.
2. The ratio test $\min\limits_{i|(\boldsymbol{\alpha}_q)_i>0} \{\beta_i/(\boldsymbol{\alpha}_q)_i\}$ determines the leaving basis variable, for $i = p$, say.
3. The pivot operation interchanging $(\mathbf{x}_N)_q$ with $(\mathbf{x}_B)_p$ in the basis yields a new $\mathbf{B}$, with a corresponding new tableau.

The new tableau is optimal for $\theta = \theta_c$. The reduced cost (now given by $d_j + \theta_c \hat{d}_j$) of the incoming nonbasic variable is zero, since the old (and new) tableau is dual-degenerate at $\theta = \theta_c$.

However, the corresponding $\hat{d}_j = \hat{d}_q^{\text{new}} > 0$ since

$$\hat{d}_q^{\text{new}} = -\frac{\hat{d}_q}{(\boldsymbol{\alpha}_q)_p} \tag{5-61}$$

(cf. Eq. (4-86)), and $\hat{d}_q < 0$, $(\boldsymbol{\alpha}_q)_p > 0$.

We can therefore increase θ further and again determine a new θ_c for which a change of basis is required.

The process terminates in one of three ways:

1. $\theta_c \geqslant \theta_{\max}$. Normal termination in which we have determined all the solutions over the prespecified range.
2. $\hat{d}_j \geqslant 0$ for all $j = 1, \ldots, (n - m)$. In this case $\theta_c = \infty$ and the current solution remains optimal for all increasing values of θ since the reduced costs never turn negative.
3. The solution goes unbounded in performing the ratio test on a basis change ($\alpha_i \leqslant 0$, $i = 1, \ldots, m$). Thus for all $\theta > \theta_c$ the solution is unbounded.

5-3-2 Parametric Variation of the Right-hand Side Vector

We may state the problem in the form

$$\begin{array}{ll} \text{Maximize} & P = \mathbf{c}^{\mathsf{T}}\mathbf{x} \\ \text{subject to} & \mathbf{A}\mathbf{x} = \mathbf{b} + \theta\hat{\mathbf{b}} \\ \text{in the range} & 0 \leqslant \theta \leqslant \theta_{\max} \\ & \mathbf{x} \geqslant \mathbf{0} \end{array} \tag{5-62}$$

Again we solve the problem in the normal way with $\theta = 0$.

Suppose now the optimal tableau is of the form

$$\begin{aligned} P &= \beta_0 + \hat{\beta}_0\theta - d_1(\mathbf{x}_N)_1 - \cdots - d_j(\mathbf{x}_N)_j - \cdots \\ \mathbf{x}_B &= \boldsymbol{\beta} + \hat{\boldsymbol{\beta}}\theta - \boldsymbol{\alpha}_1(x_N)_1 - \cdots - \boldsymbol{\alpha}_j(\mathbf{x}_N)_j - \cdots \end{aligned} \tag{5-63}$$

where $\beta_0 = \mathbf{c}_B^{\mathsf{T}}\mathbf{B}^{-1}\mathbf{b}$, $\hat{\beta}_0 = \mathbf{c}_B^{\mathsf{T}}\mathbf{B}^{-1}\hat{\mathbf{b}}$, $\hat{\boldsymbol{\beta}} = \mathbf{B}^{-1}\hat{\mathbf{b}}$, and $\mathbf{B}$ is the optimal basis. (Note that $\hat{\beta}_i$, $i = 1, \ldots, m$, need not necessarily be positive.)

Now since the tableau is optimal, we have

$$\begin{aligned} \beta_i &\geqslant 0 \qquad i = 1, \ldots, m \\ d_j &\geqslant 0 \qquad j = 1, \ldots, n - m \end{aligned} \tag{5-64}$$

As θ is increased from zero, since each $d_j \geqslant 0$, the solution remains optimal as long as it remains feasible, which is as long as

$$\beta_i + \theta\hat{\beta}_i \geqslant 0 \qquad i = 1, \ldots, m \tag{5-65}$$

We thus compute

$$\theta_c = \min_{i \mid \hat{\beta}_i < 0} \left\{ \frac{\beta_i}{-\hat{\beta}_i} \right\} \tag{5-66}$$

for $i = p$, say.

This minimum positive ratio gives the largest value of θ for which the solution remains feasible (and optimal), since for $\theta > \theta_c$, $(\mathbf{x}_B)_p$ is negative.

At $\theta = \theta_c$ we therefore perform a change of basis, using the dual-simplex method, discussed in Sec. 4-7.

1. $(\mathbf{x}_B)_p$ is made nonbasic.
2. To investigate which nonbasic variable should replace $(\mathbf{x}_B)_p$ in the basis, we examine the ratio

$$\frac{-d_j}{(\boldsymbol{\alpha}_j)_p} \qquad j = 1, \ldots, n - m \tag{5-67}$$

   Comparing this with Eq. (4-86), we see that this is the reduced cost of $(\mathbf{x}_B)_p$ in the new tableau if $(\mathbf{x}_N)_j$ replaces $(\mathbf{x}_B)_p$ in the basis.

   We have primal degeneracy when $\theta = \theta_c$, in that $(\mathbf{x}_B)_p = 0$ as well as all the nonbasic variables, so we are free to choose any nonbasic variable to pivot with, as long as the reduced costs remain nonnegative.

   The ratio test becomes

$$\min_{j \mid (\boldsymbol{\alpha}_j)_p < 0} \left\{ \frac{-d_j}{(\boldsymbol{\alpha}_j)_p} \right\} \tag{5-68}$$

   for $j = q$, say, as in the dual-simplex method.
3. The pivot operation interchanging $(\mathbf{x}_N)_q$ with $(\mathbf{x}_B)_p$ in the basis yields a new **B**, with a corresponding new tableau.

The new tableau is optimal for $\theta = \theta_c$. The current right-hand side (now given by $\boldsymbol{\beta} + \theta_c\hat{\boldsymbol{\beta}}$) is primal degenerate at $i = p$, since $(\mathbf{x}_N)_q$ enters at the zero level. However, the corresponding $\hat{\beta}_p^{\text{new}} > 0$, since

$$\hat{\beta}_p^{\text{new}} = \frac{\hat{\beta}_p}{(\boldsymbol{\alpha}_q)_p} \tag{5-69}$$

and $(\boldsymbol{\alpha}_q)_p < 0$, $\hat{\beta}_p < 0$. We can therefore increase θ further and again determine a new θ_c for which a change of basis is required.

The process terminates in one of three ways:

1. $\theta_c \geqslant \theta_{\max}$. This results in normal termination in which we have determined all the solutions over the prespecified range.
2. $\hat{\beta}_i \geqslant 0$ for all $i = 1, \ldots, m$. In this case, $\theta_c = \infty$ and the current solution remains optimal for all increasing values of θ, since the basic variables never go through zero.
3. $(\boldsymbol{\alpha}_j)_p \geqslant 0$ for all $j = 1, \ldots, n - m$. This means we cannot increase θ without making the problem infeasible (cf. Eq. 5-69). Thus, for all $\theta > \theta_c$ the solution is infeasible.

It will be evident that parametric programming on the right-hand side vector is essentially the dual of parametric programming on the cost row.

5-4 DECOMPOSITION

Decomposition is a method of exploiting the structure of a problem to reduce it from one large problem to a series of smaller problems. Different types of decomposition methods have been developed, according to the particular structure of the $\mathbf{A}$ matrix, corresponding to the constraints $\mathbf{Ax} = \mathbf{b}$. The most commonly occurring structure amenable to decomposition is the so-called *block-angular* system, which is a series of block-diagonal submatrices bound together by a common group of rows, as follows:

$$\mathbf{A} = \begin{bmatrix} \mathbf{A}_1 & \mathbf{A}_2 & \cdots & \mathbf{A}_t \\ \mathbf{B}_1 & & & \\ & \mathbf{B}_2 & & \\ & & \ddots & \\ & & & \mathbf{B}_t \end{bmatrix}$$

The problem with this structure may be written in the form

Maximize $$P = \sum_{k=1}^{t} \mathbf{c}_k^{\mathrm{T}} \mathbf{x}_k \tag{5-70}$$

subject to $$\sum_{k=1}^{t} \mathbf{A}_k \mathbf{x}_k = \mathbf{b}_0 \qquad (m_0 \text{ rows}) \tag{5-71}$$

$$\mathbf{B}_k \mathbf{x}_k = \mathbf{b}_k \qquad (m_k \text{ rows}),\ k = 1, \ldots, t \tag{5-72}$$

$$\mathbf{x}_k \geqslant 0 \qquad ,\ k = 1, \ldots, t \tag{5-73}$$

Such a structure arises naturally in multi-time period models, where the constraints $\mathbf{B}_k \mathbf{x}_k = \mathbf{b}_k$ are confined to each time period $k = 1, \ldots, t$, and the group of common rows $\sum_{k=1}^{t} \mathbf{A}_k \mathbf{x}_k = \mathbf{b}_0$ refer to constraints which hold throughout the planning horizon. The structure also arises in the modeling of hierarchical systems, where each semi-autonomous subsystem has its own constraints

$(\mathbf{B}_k\mathbf{x}_k = \mathbf{b}_k)$ as well as being subject to overriding organizational constraints $\sum_{k=1}^{t} \mathbf{A}_k\mathbf{x}_k = \mathbf{b}_0$. Clearly, if the common rows could be removed in some way, the problem could be solved by solving t smaller subproblems.

The decomposition method of Dantzig and Wolfe (1960) was developed to achieve this, and builds upon the notion of *column generation*, which means the nonbasic columns of $\mathbf{A}$ needed for the pricing operation (Step 2, Sec. 2-6) are generated only when they are required.

Another concept fundamental to the Dantzig–Wolfe decomposition principle is that, assuming the set of points $\mathbf{x}_k$ which satisfy $\mathbf{B}_k\mathbf{x}_k = \mathbf{b}_k$, $\mathbf{x}_k \geqslant 0$, is a closed and bounded set†, say S_k, then any point in the set S_k can be represented in the form

$$\mathbf{x}_k = \sum_{j=1}^{r_k} \lambda_k^j \mathbf{x}_k^j \tag{5-74}$$

where

$$\sum_{j=1}^{r_k} \lambda_k^j = 1 \qquad k = 1, \ldots, t \tag{5-75}$$

$$\lambda_k^j \geqslant 0 \qquad j = 1, \ldots, r_k \tag{5-76}$$

and $\mathbf{x}_k^j$, $j = 1, \ldots, r_k$, are the vertices (extreme points) of the set S_k.

We may therefore restate the original problem, Eqs. (5-70)–(5-73), in the form: choose, from all the solutions of Eqs. (5-72) and (5-73), those which satisfy Eq. (5-71) and maximize $\sum_{k=1}^{t} \mathbf{c}_k^{\mathrm{T}}\mathbf{x}_k$.

Using Eq. (5-74), Eq. (5-70) becomes

Maximize
$$P = \sum_{k=1}^{t} \sum_{j=1}^{r_k} (\mathbf{c}_k^{\mathrm{T}}\mathbf{x}_k^j)\lambda_k^j \tag{5-77}$$

and Eq. (5-71) becomes

$$\sum_{k=1}^{t} \sum_{j=1}^{r_k} (\mathbf{A}_k\mathbf{x}_k^j)\lambda_k^j = \mathbf{b}_0 \tag{5-78}$$

If we define

$$\mathbf{p}_k^j = \mathbf{A}_k\mathbf{x}_k^j \qquad j = 1, \ldots, r_k \tag{5-79}$$

and

$$u_k^j = \mathbf{c}_k^{\mathrm{T}}\mathbf{x}_k^j \qquad j = 1, \ldots, r_k \tag{5-80}$$

then Eqs. (5-75)–(5-78) comprise a linear-programming problem in λ_k^j, called the *master problem*,

Maximize
$$\sum_{k=1}^{t} \sum_{j=1}^{r_k} u_k^j \lambda_k^j \tag{5-81}$$

† We can always ensure we have a closed and bounded set by placing an appropriate upper bound on each element of $\mathbf{x}_k$.

subject to

$$\sum_{k=1}^{t} \sum_{j=1}^{r_k} \mathbf{p}_k^j \lambda_k^j = b_0 \tag{5-82}$$

$$\sum_{j=1}^{r_k} \lambda_k^j = 1 \qquad k = 1, \ldots, t \tag{5-83}$$

$$\lambda_k^j \geqslant 0 \qquad j = 1, \ldots, r_k, k = 1, \ldots, t \tag{5-84}$$

Note that this problem has only $m_0 + t$ rows, compared with $m_0 + \sum_{k=1}^{t} m_k$ rows of the original problem; thus there is considerable potential saving. The function of the master problem is to assign weights λ_k^j to the solutions $\mathbf{x}_k^j$ which satisfy Eq. (5-72) and (5-73).

However, the number of columns in the master problem is equal to the number of vertices in S_k for $k = 1, \ldots, t$, which could be very large. We therefore use a column-generating technique in solving the master problem, creating the columns as required.

Suppose that in solving the master problem, we have produced a pricing vector $\boldsymbol{\pi}^{\mathrm{T}}$ (Step 1, Sec. 2-6) and we now need a nonbasic column to carry out the pricing operation.

The reduced cost of the nonbasic variable λ_k^j is given by

$$d_k^j = \boldsymbol{\pi}^{\mathrm{T}} \begin{bmatrix} \mathbf{p}_k^j \\ \hline 1 \end{bmatrix} - u_k^j \tag{5-85}$$

Partitioning $\boldsymbol{\pi}^{\mathrm{T}}$ into $(\boldsymbol{\pi}_1^{\mathrm{T}} : \pi_0)$, corresponding to Eqs. (5-82) and (5-83) respectively, this becomes

$$\begin{aligned} d_k^j &= \boldsymbol{\pi}_1^{\mathrm{T}} \mathbf{p}_k^j + \pi_0 - u_k^j \\ &= \boldsymbol{\pi}_1^{\mathrm{T}} A_k \mathbf{x}_k^j + \pi_0 - \mathbf{c}_k^{\mathrm{T}} \mathbf{x}_k^j \\ &= -(\mathbf{c}_k - \boldsymbol{\pi}_1^{\mathrm{T}} A_k)\mathbf{x}_k^j + \pi_0 \end{aligned} \tag{5-86}$$

Finding the most negative reduced cost is equivalent to solving the linear-programming problem, called the kth *subproblem*,

Maximize	$(\mathbf{c}_k - \boldsymbol{\pi}_1^{\mathrm{T}} \mathbf{A}_k)\mathbf{x}_k$	(5-87)
subject to	$\mathbf{B}_k \mathbf{x}_k = \mathbf{b}_k$	(5-88)
	$\mathbf{x}_k \geqslant 0$	(5-89)

The solution, $\mathbf{x}_k^q$ say, will be an extreme point of S_k and will correspond to the most negative reduced cost. We will have generated the required column

$$\begin{bmatrix} \mathbf{p}_k^q \\ \hline 1 \end{bmatrix} = \begin{bmatrix} \mathbf{A}_k \mathbf{x}_k^q \\ 1 \end{bmatrix} \tag{5-90}$$

which is now available to enter the basis of the master problem, and the corresponding cost coefficient is given by $u_k^q = \mathbf{c}_k^{\mathrm{T}} \mathbf{x}_k^q$.

This process is repeated for all $k = 1, \ldots, t$, and the solution whose reduced cost is most negative out of all the t subproblems has its corresponding column enter the basis, and the remaining simplex steps 4, 5, and 6 (Sec. 2-6) of the master problem continued with, thus giving rise to a new pricing vector in the next iteration.

As an alternative to examining each subproblem individually, it is possible to aggregate them by defining

$$\mathbf{p}_j = \sum_{k=1}^{t} \mathbf{A}_k \mathbf{x}_k^j \tag{5-91}$$

and

$$u_j = \sum_{k=1}^{t} \mathbf{c}_k^\top \mathbf{x}_k^j \tag{5-92}$$

The master problem then has $m_0 + 1$ rows instead of $m_0 + t$, and the single subproblem

Maximize
$$\sum_{k=1}^{t} (\mathbf{c}_k^\top - \boldsymbol{\pi}_1^\top \mathbf{A}_k)\mathbf{x}_k \tag{5-93}$$

subject to
$$\mathbf{B}_k \mathbf{x}_k = \mathbf{b}_k \qquad k = 1, \ldots, t \tag{5-94}$$

$$\mathbf{x}_k \geqslant 0 \tag{5-95}$$

This can be solved as the sum of the t individual subproblems, and their individual reduced costs aggregated.

The first alternative would normally be recommended as it has more flexibility; the price paid for this flexibility is an extra $(t - 1)$ rows in the master problem. Note, however, that the structure of the t rows is amenable to special treatment, using generalized upper bounding.

An interesting economic interpretation can be made of the decomposition process. Note that in solving the kth subproblem, the master problem specifies a set of *prices*, $\boldsymbol{\pi}_1$, which represent the rate of change of objective function P with respect to the right-hand side vector of the common resources, $\mathbf{b}_0$. We can draw the analogy of the master problem representing central management, who specify a set of internal prices, within which each division, k, responds with a set of proposals $\mathbf{p}_k^j = \mathbf{A}_j \mathbf{x}_k^j$ for utilization of the common resources, obtained by solving the kth subproblem. The problem of central management is to find the optimal combination, λ_k^j, of these proposals.

The problem of the kth division is to adjust $\mathbf{x}_k$ to maximize their profit, using $\boldsymbol{\pi}_1$ as their price per unit consumption of the common resources. Thus their objective function is $\mathbf{c}_k^\top \mathbf{x}_k$ less the cost of their consumption of the common resources, $\boldsymbol{\pi}_1^\top(\mathbf{A}_k \mathbf{x}_k)$ (cf. Eq. (5-87)).

The next iteration of the master problem will generate a new set of transfer prices for the divisions.

CHAPTER

SIX

SIMPLEX-BASED NONLINEAR AND INTEGER PROGRAMMING

6-1 SEPARABLE PROGRAMMING

This procedure, developed by Miller (1963), was devised to solve the following problem:

Minimize $$f^0(\mathbf{x}) = \sum_{j=1}^{n} f_j^0(x_j) \tag{6-1}$$

subject to $$f^i(\mathbf{x}) = \sum_{j=1}^{n} f_j^i(x_j) \leqslant b_i \qquad i = 1, \ldots, m \tag{6-2}$$

$$x_j \geqslant 0 \qquad j = 1, \ldots, n \tag{6-3}$$

The objective function and the constraints can each be *separated* into the sum of individual nonlinear functions of individual variables.

While this may appear to be unduly restrictive, it is often possible to get nonlinear functions into this form by a suitable transformation of variables; an example which occurs quite commonly in practice is the product of two variables x_1x_2. By using the equation

$$x_1x_2 = \tfrac{1}{4}(x_1 + x_2)^2 - \tfrac{1}{4}(x_1 - x_2)^2 \tag{6-4}$$

and defining $$x_3 = (x_1 + x_2) \tag{6-5}$$

$$x_4 = (x_1 - x_2) \tag{6-6}$$

we can replace the expression $x_1 x_2$ wherever it occurs by $\frac{1}{4}x_3^2 - \frac{1}{4}x_4^2$ (which are functions of the individual variables x_3 and x_4) and add the two equality constraints

$$x_1 + x_2 - x_3 = 0$$
$$x_1 - x_2 - x_4 = 0$$

Each individual function is approximated by a piecewise linearization, as shown in Fig. 6-1.

The nonlinear function is approximated by a series of straight lines joining the grid points $(a_1, b_1), \ldots, (a_k, b_k)$ where a_k is the x coordinate and b_k is the $f(x)$ coordinate.

We introduce k new nonnegative *special variables*, $\lambda_1, \ldots, \lambda_k$, and introduce the equations

$$\lambda_1 + \lambda_2 + \cdots + \lambda_k = 1 \tag{6-7}$$
$$a_1\lambda_1 + a_2\lambda_2 + \cdots + a_k\lambda_k = x \tag{6-8}$$
$$b_1\lambda_1 + b_2\lambda_2 + \cdots + b_k\lambda_k = f(x) \tag{6-9}$$

The left-hand side of Eq. (6-9) replaces $f(x)$ wherever it occurs, and Eqs. (6-7) and (6-8) become additional constraints. By ensuring that no more than two contiguous λ's are nonzero, the special variables become interpolating parameters between adjacent grid points.

Each nonlinear function $f_j^i(x_j)$, $i = 0, 1, \ldots, m$, $j = 1, \ldots, n$, requires its own set of special variables

$$S_j^i = (\lambda_1, \ldots, \lambda_k)$$

The number of grid points k need not be the same for each function, and a certain amount of art is required in selecting the number and location of these points. Beale (1968) recommends no more than six points for each function, on the grounds that the rate of convergence is slow for higher numbers.

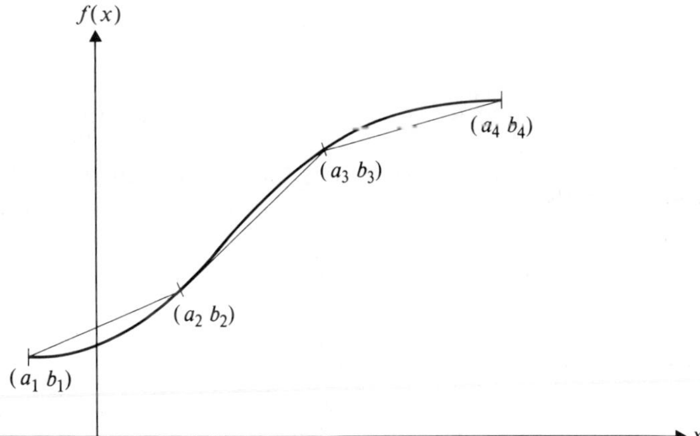

Figure 6-1 Piecewise linearization of a nonlinear function.

Example Consider the problem

Maximize x_3

subject to $x_1^2 + x_2^2 + x_3^2 = 1$

This could be rephrased as

Maximize x_3

subject to

$$x_4 + x_5 + x_6 = 1$$
$$x_1^2 = x_4$$
$$x_2^2 = x_5$$
$$x_3^2 = x_6$$

Consider the graph of $f(x_j) = x_j^2, j = 1, 2, 3$, and divide into four grid points

$$(a_1, b_1) = (0.0, 0.0)$$
$$(a_2, b_2) = (0.4, 0.16)$$
$$(a_3, b_3) = (0.7, 0.49)$$
$$(a_4, b_4) = (1.0, 1.0)$$

The problem becomes, in separable-programming format:

Maximize x_3

subject to

$$x_4 + x_5 + x_6 = 1$$
$$\lambda_1 + \lambda_2 + \lambda_3 + \lambda_4 = 1$$
$$0.4\lambda_2 + 0.7\lambda_3 + \lambda_4 = x_1$$
$$0.16\lambda_2 + 0.49\lambda_3 + \lambda_4 = x_4$$
$$\alpha_1 + \alpha_2 + \alpha_3 + \alpha_4 = 1$$
$$0.4\alpha_2 + 0.7\alpha_3 + \alpha_4 = x_2$$
$$0.16\alpha_2 + 0.49\alpha_3 + \alpha_4 = x_5$$
$$\beta_1 + \beta_2 + \beta_3 + \beta_4 = 1$$
$$0.4\beta_2 + 0.7\beta_3 + \beta_4 = x_3$$
$$0.16\beta_2 + 0.49\beta_3 + \beta_4 = x_6$$

In solving the derived problem, we must modify Step 2 of the simplex method (Sec. 2-6) so that no more than two contiguous special variables are basic at any one time.

For each set S_j^i of special variables:

1. If *no* element in S_j^i is basic, then we can consider all elements for pricing.
2. If only *one* element in S_j^i is basic, then we can consider only the variable immediately preceding it and the variable immediately following it for pricing.
3. If two elements of S_j^i are basic, then no other elements in S_j^i are considered for pricing.

6-2 MAP

MAP is an acronym for *method of approximate programming*. The method was developed by Griffith and Stewart (1961). As its name implies, it involves making a linear approximation to the nonlinear function.

For this section and the next, we will need to use a result from calculus, namely Taylor's theorem. If the function $f(\mathbf{x})$ is twice-continuously differentiable for all $\mathbf{x} \in E^n$, we have

$$f(\mathbf{x} + \Delta\mathbf{x}) = f(\mathbf{x}) + \mathbf{g}(\mathbf{x})^\top \Delta\mathbf{x} + \tfrac{1}{2}\Delta\mathbf{x}^\top \mathbf{H}(\mathbf{x} + \theta\,\Delta\mathbf{x})\,\Delta\mathbf{x} \qquad 0 < \theta < 1 \tag{6-10}$$

where $$\mathbf{g}(\mathbf{x}) = \frac{\partial f(\mathbf{x})}{\partial \mathbf{x}}$$

is the vector of first partial derivatives of f with respect to $\mathbf{x}$, $\partial f/\partial x_i$, $i = 1, \ldots, n$, and

$$\mathbf{H}(\mathbf{x} + \theta\,\Delta\mathbf{x}) = \left[\frac{\partial^2 f}{\partial x_i \partial x_j}\right]$$

is the so-called *Hessian* matrix of second partial derivatives evaluated at some point between $\mathbf{x}$ and $\mathbf{x} + \Delta\mathbf{x}$. The ijth element of this $n \times n$ matrix is $\partial^2 f/\partial x_i \partial x_j$.

In practice, usually only some of the variables are nonlinear. If we reorder the variables so that the first, say, r variables are nonlinear, then the problem may be stated in the form

Minimize $$P = f^0(x_1, \ldots, x_r) + \sum_{j=r+1}^{n} c_j x_j \tag{6-11}$$

subject to $$f^i(x_1, \ldots, x_r) + \sum_{j=r+1}^{n} a_{ij} x_j = b_i \qquad i = 1, \ldots, m \tag{6-12}$$

$$l_j \leqslant x_j \leqslant u_j \qquad j = 1, \ldots, r \tag{6-13}$$

$$x_j \geqslant 0 \qquad j = r + 1, \ldots, n \tag{6-14}$$

For largely historical reasons, it is customary to consider minimizing nonlinear functions, but there is no loss in generality since minimizing $f(\mathbf{x})$ is equivalent to maximizing $-f(\mathbf{x})$.

The linear approximation used in MAP comes from considering the first two terms in Taylor's series,

$$f^i(x_1, \ldots, x_r) \approx f^i(x_1^0, \ldots, x_r^0) + \sum_{j=1}^{r} g_j^i\,\Delta x_j \tag{6-15}$$

where $$g_j^i = \frac{\partial f^i}{\partial x_j}(x_1^0, \ldots, x_r^0) \tag{6-16}$$

and $$\Delta x_j = x_j - x_j^0 \tag{6-17}$$

This is equivalent to approximating the function by its tangent at the point $(x_1^0, \ldots, x_r^0)$. A one-dimensional example is illustrated in Fig. 6-2.

Substituting Eq. (6-15) into Eqs. (6-11) and (6-12), the linearized problem becomes

Minimize

$$P = f^0(x_1^0, \ldots, x_r^0) + \sum_{j=1}^{r} g_j^0 \, \Delta x_j + \sum_{j=r+1}^{n} c_j x_j \tag{6-18}$$

subject to

$$f^i(x_1^0, \ldots, x_r^0) + \sum_{j=1}^{r} g_j^i \, \Delta x_j + \sum_{j=r+1}^{n} a_{ij} x_j = b_i \qquad i = 1, \ldots, m \tag{6-19}$$

$$l_j - x_j^0 \leqslant \Delta x_j \leqslant u_j - x_j^0 \qquad j = 1, \ldots, r \tag{6-20}$$

$$x_j \geqslant 0 \qquad j = r+1, \ldots, n \tag{6-21}$$

The first term in Eqs. (6-18) and (6-19) is constant, and the problem is now a linear programming problem in the variables $\Delta x_j, j = 1, \ldots, r; x_j, j = r+1, \ldots, n$.

In practice, it is usual to limit the size of the step Δx_j, $j = 1, \ldots, r$, since too large a step will make the linear approximation grossly invalid. The bounds on Δx_j become

$$l_j' \leqslant \Delta x_j \leqslant u_j' \qquad j = 1, \ldots, r \tag{6-22}$$

where

$$l_j' = \max \{l_j - x_j^0, -m_j\} \tag{6-23}$$

$$u_j' = \min \{u_j - x_j^0, +m_j\} \tag{6-24}$$

and m_j is the move limit on Δx_j.

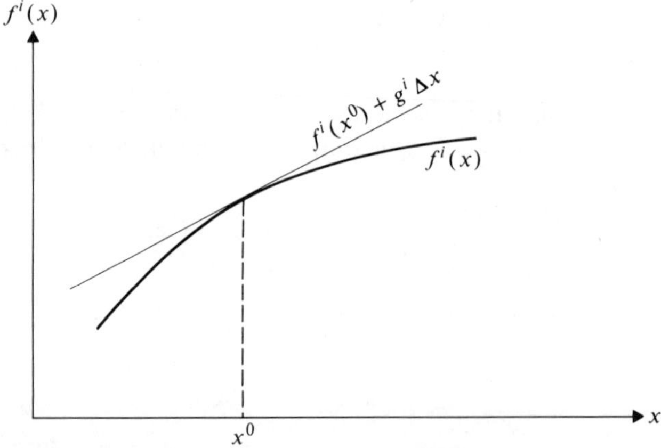

Figure 6-2 Linearization of $f^i(x)$ at x^0.

The procedure can be summarized as follows:

1. Start with an estimate $(x_1^0, \ldots, x_r^0)$ of the optimum values of $(x_1, \ldots, x_r)$ and linearize the nonlinear functions $f^i(x_1, \ldots, x_r)$, $i = 0, 1, \ldots, m$, about $(x_1^0, \ldots, x_r^0)$.
2. Solve the linear programming problem, Eqs. (6-18), (6-19), (6-21), (6-22), yielding $(\Delta x_1, \ldots, \Delta x_r, x_{r+1}, \ldots, x_n)$.
3. The new estimate becomes $(x_1^0 + \Delta x_1, \ldots, x_r^0 + \Delta x_r)$. Repeat from Step 1.

The criterion for stopping the procedure is when there is no appreciable movement from one iteration to the next.

The method is adequate as long as the functions are nearly linear, otherwise extensive "fine tuning" of the move limits m_j is required to strike a delicate balance between rate of convergence and accuracy of the approximation. However, if many similar problems are to be solved on a production basis, the effort may well be worth while, and the method has enjoyed widespread usage. For example, Buzby (1974) reports solving problems with $r = 200$, $n = 10{,}000$ and $m = 2{,}200$ on a routine basis.

6-3 MINOS

MINOS is the name given to a procedure developed by Murtagh and Saunders (1978) for solving large-scale nonlinear programming problems with linear constraints. Unlike the methods discussed in the previous two sections, it does not make any approximation to $f(\mathbf{x})$ but works with the nonlinear objective function directly.

The problem may be stated in the form

$$\text{Minimize} \qquad P = f(x_1, \ldots, x_r) + \sum_{j=r+1}^{n} c_j x_j \qquad (6\text{-}25)$$

$$\text{subject to} \qquad \mathbf{Ax} = \mathbf{b} \qquad (6\text{-}26)$$

$$\mathbf{l} \leqslant \mathbf{x} \leqslant \mathbf{u} \qquad (6\text{-}27)$$

Since the objective function is nonlinear, we cannot assume that exactly m variables will be basic at an optimum solution. We thus introduce the notion of *superbasic* variables and partition the set of constraints, Eq. (6-26), as follows

$$\mathbf{Ax} = \overset{\qquad m \qquad\quad s \qquad\quad n-m-s}{\left[\begin{array}{c|c|c} \mathbf{B} & \mathbf{S} & \mathbf{N} \end{array}\right]}_{\text{basics} \quad \text{superbasics} \quad \text{nonbasics}} \begin{bmatrix} \mathbf{x}_B \\ \hline \mathbf{x}_S \\ \hline \mathbf{x}_N \end{bmatrix} = \mathbf{b} \qquad (6\text{-}28)$$

The matrix $\mathbf{B}$ is square and nonsingular and corresponds to the usual basis matrix of the simplex method; $\mathbf{S}$ is $m \times s$ with $0 \leqslant s \leqslant n - m$ and the associated variables $\mathbf{x}_S$ are called the *superbasics* as shown. The nonbasic variables $\mathbf{x}_N$ are at one or other of their bounds, as in the simplex method.

It is easily shown that an optimal solution to the problem, Eqs. (6-25)–(6-27), exists for which $s \leqslant r$. (For example, set all the nonlinear variables to their optimal values.) We thus keep s as small as possible and proceed in an analogous fashion to the simplex method.

There are two desirable properties of the step $\Delta\mathbf{x}$ from $\mathbf{x}$ to $\mathbf{x} + \Delta\mathbf{x}$:

Property 1

$$\left[\begin{array}{c:c:c} \mathbf{B} & \mathbf{S} & \mathbf{N} \\ \hdashline 0 & 0 & \mathbf{I} \end{array}\right] \begin{bmatrix} \Delta\mathbf{x}_B \\ \hline \Delta\mathbf{x}_S \\ \hline \Delta\mathbf{x}_N \end{bmatrix} = \mathbf{0} \tag{6-29}$$

This merely states that the new point $\mathbf{x} + \Delta\mathbf{x}$ should be feasible, i.e., satisfy Eqs. (6-26), (6-27).

Property 2

$$\begin{bmatrix} \mathbf{g}_B \\ \hline \mathbf{g}_S \\ \hline \mathbf{g}_N \end{bmatrix} + \mathbf{H} \begin{bmatrix} \Delta\mathbf{x}_B \\ \hline \Delta\mathbf{x}_S \\ \hline \Delta\mathbf{x}_N \end{bmatrix} = \left[\begin{array}{c:c} \mathbf{B}^\top & 0 \\ \hline \mathbf{S}^\top & 0 \\ \hline \mathbf{N}^\top & \mathbf{I} \end{array}\right] \begin{bmatrix} \boldsymbol{\mu} \\ \boldsymbol{\lambda} \end{bmatrix} \tag{6-30}$$

(The vectors $\mathbf{g} = \mathbf{g}(\mathbf{x})$ and $\Delta\mathbf{x}$ have been partitioned corresponding to the partitioning of $\mathbf{x}$.)

The gradient of the objective function at $\mathbf{x} + \Delta\mathbf{x}$ is given by the left-hand side of Eq. (6-30) if $\mathbf{x}$ is close enough to $\mathbf{x} + \Delta\mathbf{x}$ for a quadratic approximation to be adequate. For $\mathbf{x} + \Delta\mathbf{x}$ to be locally optimal with respect to the current set of active constraints, we require that the gradient of the objective function is orthogonal to the surface formed by the active constraints. This means that there is no component of the gradient vector along the surface, so no further progress can be made. For the gradient vector to be orthogonal to the constraint surface, it must be a linear combination of the constraint normals, given by the right-hand side of Eq. (6-30). $\boldsymbol{\lambda}$ and $\boldsymbol{\mu}$ are called *Lagrange multipliers*.

A further necessary condition for optimality requires that the negative gradient (steepest descent) direction be pointing orthogonally *out* of the feasible region, otherwise further progress could be made by coming off an inequality constraint boundary surface. For our problem, Eqs. (6-25)–(6-27), we thus require that $\lambda_j \leqslant 0$ if $(\mathbf{x}_N)_j$ is at its upper bound, or $\lambda_j \geqslant 0$ if $(\mathbf{x}_N)_j$ is at its lower bound, $j = 1, \ldots, n - m - s$.

From Eq. (6-29) it is readily apparent that

$$\Delta \mathbf{x} = \begin{bmatrix} -\mathbf{W} \\ \hline \mathbf{I} \\ \hline \mathbf{0} \end{bmatrix} \Delta \mathbf{x}_S \tag{6-31}$$

where

$$\mathbf{W} = \mathbf{B}^{-1}\mathbf{S} \tag{6-32}$$

This means that we can work with the vector $\Delta \mathbf{x}_S$ (of dimension s, which is kept small) and generate the rest of $\Delta \mathbf{x}$ as required.

W is never actually computed; just a factorization of **B**. Equation (6-30) simplifies when premultiplied by the matrix

$$\begin{bmatrix} \mathbf{I} & \mathbf{0} & \mathbf{0} \\ -\mathbf{W}^{\mathsf{T}} & \mathbf{I} & \mathbf{0} \\ \mathbf{0} & \mathbf{0} & \mathbf{I} \end{bmatrix} \tag{6-33}$$

Considering the first row partitioning, we get

$$\mathbf{B}^{\mathsf{T}}\boldsymbol{\mu} = \mathbf{g}_B + [\mathbf{I} \quad \mathbf{0} \quad \mathbf{0}]\,\mathbf{H} \begin{bmatrix} -\mathbf{W} \\ \mathbf{I} \\ \mathbf{0} \end{bmatrix} \Delta \mathbf{x}_S \tag{6-34}$$

When $\Delta \mathbf{x}_S = \mathbf{0}$ (which will mean no further progress can be made in the current set of active constraints), Eq. (6-34) becomes

$$\mathbf{B}^{\mathsf{T}}\boldsymbol{\mu} = \mathbf{g}_B \tag{6-35}$$

in which case $\boldsymbol{\mu}$ is analogous to the pricing vector $\boldsymbol{\pi}$ of the simplex method (cf. Eq. (2-30)). We shall denote the solution of Eq. (6-35) by $\boldsymbol{\pi}$.

Considering the third row partitioning, we get

$$\boldsymbol{\lambda} = \mathbf{g}_N - \mathbf{N}^{\mathsf{T}}\boldsymbol{\mu} + (\mathbf{0} \quad \mathbf{0} \quad \mathbf{I})\,\mathbf{H} \begin{bmatrix} -\mathbf{W} \\ \mathbf{I} \\ \mathbf{0} \end{bmatrix} \Delta \mathbf{x}_S \tag{6-36}$$

and again when $\Delta \mathbf{x}_S = \mathbf{0}$, Eq. (6-36) becomes

$$\boldsymbol{\lambda} = \mathbf{g}_N - \mathbf{N}^{\mathsf{T}}\boldsymbol{\pi} \tag{6-37}$$

which is analogous to the vector of reduced costs d_j in the simplex method (see Eqs. (2-31), (2-32)).

Considering the second row partitioning of Eq. (6-30), following premultiplication by the matrix in Eq. (6-33), we get

$$(-\mathbf{W}^{\mathsf{T}} \quad \mathbf{I} \quad \mathbf{0})\,\mathbf{H}\begin{bmatrix}-\mathbf{W}\\ \mathbf{I}\\ \mathbf{0}\end{bmatrix}\Delta\mathbf{x}_S = -\mathbf{h} \tag{6-38}$$

where
$$\mathbf{h} = (-\mathbf{W}^{\mathsf{T}} \quad \mathbf{I} \quad \mathbf{0})\,\mathbf{g} = \mathbf{g}_S - \mathbf{S}^{\mathsf{T}}\pi \tag{6-39}$$

Thus we have a Newton-like step in the independent variables $\Delta\mathbf{x}_S$. $\mathbf{h}$ could be regarded as a type of "reduced" gradient, and the matrix

$$(-\mathbf{W}^{\mathsf{T}} \quad \mathbf{I} \quad \mathbf{0})\,\mathbf{H}\begin{bmatrix}-\mathbf{W}\\ \mathbf{I}\\ \mathbf{0}\end{bmatrix} \tag{6-40}$$

could be regarded as a type of "reduced" Hessian matrix. In practice, we form an approximation to Eq. (6-40) of the form $\mathbf{R}^{\mathsf{T}}\mathbf{R}$. $\mathbf{R}$ is upper triangular so that the step $\Delta\mathbf{x}_S$ given by

$$\mathbf{R}^{\mathsf{T}}\mathbf{R}\,\Delta\mathbf{x}_S = -\mathbf{h} \tag{6-41}$$

can be calculated using simple forward and back substitution, in the same way as the **LU** factorization of **B** discussed in Sec. 3-4-2.

6-3-1 Summary of Procedure

Assume we have the following:

(*a*) A feasible vector $\mathbf{x} = [\mathbf{x}_B \quad \mathbf{x}_S \quad \mathbf{x}_N]^{\mathsf{T}}$ satisfying $[\mathbf{B}{:}\mathbf{S}{:}\mathbf{N}]\mathbf{x} = \mathbf{b}$, $\mathbf{l} \leqslant \mathbf{x} \leqslant \mathbf{u}$.
(*b*) The corresponding function value $f(\mathbf{x})$ and gradient vector $\mathbf{g}(\mathbf{x}) = [\mathbf{g}_B \quad \mathbf{g}_S \quad \mathbf{g}_N]^{\mathsf{T}}$.
(*c*) The number of superbasic variables; s $(0 \leqslant s \leqslant n - m)$.
(*d*) A factorization, **LU**, of the $m \times m$ basis matrix **B**.
(*e*) A factorization, $\mathbf{R}^{\mathsf{T}}\mathbf{R}$, of a quasi-Newton approximation to the $s \times s$ matrix

$$[\mathbf{W}^{\mathsf{T}} \quad \mathbf{I} \quad \mathbf{0}]\,\mathbf{H}\begin{bmatrix}-\mathbf{W}\\ \mathbf{I}\\ \mathbf{0}\end{bmatrix}$$

(Note that this matrix is never actually computed.)
(*f*) A vector π satisfying $\mathbf{B}^{\mathsf{T}}\pi = \mathbf{g}_B$.
(*g*) The reduced gradient vector $\mathbf{h} = \mathbf{g}_S - \mathbf{S}^{\mathsf{T}}\pi$.
(*h*) Small positive convergence tolerances TOLRG and TOLDJ.

Step 1 (Test for convergence in the current subspace). If $\|\mathbf{h}\| >$ TOLRG, go to Step 3.

Step 2 ("PRICE", i.e., estimate Lagrange multipliers, add one superbasic).

(*a*) Calculate $\boldsymbol{\lambda} = \mathbf{g}_N - \mathbf{N}^{\mathsf{T}}\boldsymbol{\pi}$.
(*b*) Select $\lambda_{q1} < -\text{TOLDJ}$ ($\lambda_{q2} > +\text{TOLDJ}$), the largest elements of $\boldsymbol{\lambda}$ corresponding to variables at their lower (upper) bound. If none, STOP; the necessary conditions for an optimal solution are satisfied.
(*c*) Otherwise,

(i) Choose $q = q_1$ or $q = q_2$ corresponding to

$$|\lambda_q| = \max |\lambda_{q1}|, |\lambda_{q2}|;$$

(ii) add $\mathbf{a}_q$ as a new column of $\mathbf{S}$;
(iii) add λ_q as a new element of $\mathbf{h}$;
(iv) add a suitable new column to $\mathbf{R}$.

(*d*) Increase s by 1.

Step 3 (Compute direction of search, $\mathbf{p}$)

(*a*) Solve $\mathbf{R}^{\mathsf{T}}\mathbf{R}\mathbf{p}_S = -\mathbf{h}$.
(*b*) Solve $\mathbf{L}\mathbf{U}\mathbf{p}_B = -\mathbf{S}\mathbf{p}_S$.
(*c*) Set $\mathbf{p} = \begin{bmatrix} \mathbf{p}_B \\ \mathbf{p}_S \\ \mathbf{0} \end{bmatrix}$.

Step 4 (Ratio test, "CHUZR")

(*a*) Find $\alpha_{\max} \geqslant 0$, the greatest value of α for which $\mathbf{x} + \alpha\mathbf{p}$ is feasible.
(*b*) If $\alpha_{\max} = 0$, go to Step 7.

Step 5 (Linesearch)

(*a*) Find α, an approximation to α^*, where

$$f(\mathbf{x} + \alpha^*\mathbf{p}) = \min_{0 < \theta \leqslant \alpha\max} f(\mathbf{x} + \theta\mathbf{p}).$$

Set $\Delta\mathbf{x} = \alpha\mathbf{p}$.
(*b*) Change $\mathbf{x}$ to $\mathbf{x} + \Delta\mathbf{x}$ and set f and $\mathbf{g}$ to their values at the new $\mathbf{x}$.

Step 6 (Compute reduced gradient, $\mathbf{h}$)

(*a*) Solve $\mathbf{U}^{\mathsf{T}}\mathbf{L}^{\mathsf{T}}\boldsymbol{\pi} = \mathbf{g}_B$.
(*b*) Compute the new reduced gradient, $\bar{\mathbf{h}} = \mathbf{g}_S - \mathbf{S}^{\mathsf{T}}\boldsymbol{\pi}$.

(*c*) Modify $\mathbf{R}$ to reflect some quasi-Newton recursion on $\mathbf{R}^{\top}\mathbf{R}$, using α, $\mathbf{p}_S$ and the change in reduced gradient, $\bar{\mathbf{h}} - \mathbf{h}$.
(*d*) Set $\mathbf{h} = \bar{\mathbf{h}}$.
(*e*) If $\alpha < \alpha_{max}$, go to Step 1. No new constraint was encountered so we remain in the current subspace.

Step 7 (Change basis if necessary; delete one superbasic). Here $\alpha = \alpha_{max}$, and for some p $(0 < p \leqslant m + s)$, a variable corresponding to the pth column of $[\mathbf{B} \quad \mathbf{S}]$ has reached one of its bounds.

(*a*) If a *basic* variable hits its bound $(0 < p \leqslant m)$:

(i) interchange the pth and qth columns of

$$\begin{bmatrix} \mathbf{B} \\ \mathbf{x}_B^{\top} \end{bmatrix} \quad \text{and} \quad \begin{bmatrix} \mathbf{S} \\ \mathbf{x}_S^{\top} \end{bmatrix}$$

respectively, where q is chosen to keep $\mathbf{B}$ nonsingular (this requires a vector $\boldsymbol{\pi}_p$ which satisfies $\mathbf{U}^{\top}\mathbf{L}^{\top}\boldsymbol{\pi}_p = \mathbf{e}_p$; q is chosen so that $|\boldsymbol{\pi}_p^{\top}\mathbf{S}\mathbf{e}_q| \neq 0$);
(ii) modify $\mathbf{L}$, $\mathbf{U}$, $\mathbf{R}$, and $\boldsymbol{\pi}$ to reflect this change in $\mathbf{B}$;
(iii) compute the new reduced gradient $\mathbf{h} = \mathbf{g}_S - \mathbf{S}^{\top}\boldsymbol{\pi}$;
(iv) go to (*c*).

(*b*) Otherwise, a *superbasic* variable hits its bound $(m < p \leqslant m + s)$. Define $q = p - m$.
(*c*) Make the qth variable in $\mathbf{S}$ nonbasic at the appropriate bound, thus:

(i) delete the qth columns of

$$\begin{bmatrix} \mathbf{S} \\ \mathbf{x}_S^{\top} \end{bmatrix} \quad \text{and} \quad \begin{bmatrix} \mathbf{R} \\ \mathbf{h}^{\top} \end{bmatrix};$$

(ii) restore $\mathbf{R}$ to triangular form.

(*d*) Decrease s by 1 and go to Step 1.

The work involved in one pass through the above procedure is roughly equivalent to:

(*a*) one iteration of the revised simplex method on a linear program of dimensions $m \times n$; plus
(*b*) one iteration of a quasi-Newton algorithm on an unconstrained optimization problem of dimension s.

Note that the PRICE operation (Step 2) is performed only when $\|\mathbf{h}\|$ is sufficiently small, which means an average of about once every five iterations.

This is a typical frequency in commercial LP systems using multiple pricing. The extra work involved in the quasi-Newton steps is somewhat offset by the fact that a basis change (Step 7(*a*)) occurs only occasionally, so the growth of nonzeros in the **LU** factors of **B** is minimal. Thus, if s is of reasonable size and if $f(\mathbf{x})$ and $\mathbf{g}(\mathbf{x})$ are inexpensive to compute, iterations on a large problem will proceed at about the same time per iteration as if the problem were entirely linear.

The method is most useful when there are significant nonlinearities in the objective function. It is generally recognized that some account of second-order terms is required for such functions, and that simple sequential linearization of the objective function may produce convergence to the wrong point, or no convergence at all; MINOS overcomes these difficulties.

6-3-2 Extension to Nonlinear Constraints

One approach to nonlinearly constrained problems is to solve a sequence of linearly constrained problems, in which the nonlinearities are adjoined to the objective function with Lagrange multiplier estimates.

Let us assume that the nonlinearly constrained problem can be expressed in the following standard form:

$$\text{Minimize} \qquad f^0(\mathbf{x}^N) + \mathbf{c}^\top \mathbf{x} \tag{6-42}$$

$$\text{subject to} \qquad \mathbf{f}(\mathbf{x}^N) + \mathbf{A}_1\mathbf{x}^L = \mathbf{b}_1 \qquad (m_1 \text{ rows}) \tag{6-43}$$

$$\mathbf{A}_2\mathbf{x}^N + \mathbf{A}_3\mathbf{x}^L = \mathbf{b}_2 \qquad (m_2 \text{ rows}) \tag{6-44}$$

$$\mathbf{l} \leqslant \mathbf{x} \leqslant \mathbf{u} \qquad m = m_1 + m_2 \tag{6-45}$$

$$\text{where} \qquad \mathbf{x} = \begin{bmatrix} \mathbf{x}^N \\ \mathbf{x}^L \end{bmatrix} \qquad \text{and} \qquad \mathbf{f}(\mathbf{x}^N) = \begin{bmatrix} f^1(\mathbf{x}^N) \\ \vdots \\ f^{m_1}(\mathbf{x}^N) \end{bmatrix}$$

The first n_1 variables are nonlinear ($\mathbf{x}^N$), and the first m_1 constraints have nonlinear portions given by $f^i(\mathbf{x}^N)$, $i = 1, \ldots, m_1$. There may be purely linear constraints in the variables, given by Eq. (6-44), and general equality and inequality constraints are accommodated by placing suitable upper and lower bounds on the slack variables associated with each row.

Once again, this form emphasizes the fact that in most applications the number of nonlinear variables and constraints form a relatively small proportion of the total.

The solution process consists of a sequence of "major iterations," each one consisting of a linearization of the nonlinear constraints at some base point $\mathbf{x}_k$, corresponding to a first-order Taylor's series approximation

$$f^i(\mathbf{x}^N) = f^i(\mathbf{x}_k^N) + \frac{\partial f^i}{\partial \mathbf{x}}(\mathbf{x}_k^N)^\top(\mathbf{x}^N - \mathbf{x}_k^N) + O(\| \mathbf{x}^N - \mathbf{x}_k^N \|^2) \tag{6-46}$$

We thus define

$$\tilde{\mathbf{f}}(\mathbf{x}_k^N, \mathbf{x}^N) = \mathbf{f}(\mathbf{x}_k^N) + \mathbf{J}_k(\mathbf{x}^N - \mathbf{x}_k^N) \tag{6-47}$$

where
$$\mathbf{J}_k = \left[\frac{\partial \mathbf{f}}{\partial x_1^N}(\mathbf{x}_k^N), \ldots, \frac{\partial \mathbf{f}}{\partial x_{n_1}^N}(\mathbf{x}_k^N) \right]$$

so that $[\mathbf{f}(\mathbf{x}^N) - \tilde{\mathbf{f}}(\mathbf{x}_k^N, \mathbf{x}^N)]$ consists of the higher-order (*nonlinear*) terms in the Taylor's series expansion about the point $\mathbf{x}_k^N$. $\mathbf{J}_k$ is the $(m_1 \times n_1)$ jacobian matrix of first partial derivatives of the nonlinear constraint functions, whose *ij*th element is

$$\frac{\partial f^i}{\partial x_j}(\mathbf{x}_k^N)$$

At each major iteration the following linearly constrained problem is solved,

Minimize
$$L(\mathbf{x}_k, \boldsymbol{\lambda}_k, \mathbf{x}) = f^0(\mathbf{x}^N) - \boldsymbol{\lambda}_k^{\top}[\mathbf{f}(\mathbf{x}^N) - \tilde{\mathbf{f}}(\mathbf{x}_k^N, \mathbf{x}^N)] + \mathbf{c}^{\top}\mathbf{x} + \tfrac{1}{2}\rho[\mathbf{f} - \tilde{\mathbf{f}}]^{\top}[\mathbf{f} - \tilde{\mathbf{f}}] \tag{6-48}$$

subject to
$$\tilde{\mathbf{f}}(\mathbf{x}_k^N, \mathbf{x}^N) + \mathbf{A}_1\mathbf{x}^L = \mathbf{b}_1 \tag{6-49}$$
$$\mathbf{A}_2\mathbf{x}^N + \mathbf{A}_3\mathbf{x}^L = \mathbf{b}_2 \tag{6-50}$$
$$\mathbf{l} \leqslant \mathbf{x} \leqslant \mathbf{u} \tag{6-51}$$

The objective function, Eq. (6-48), is an *augmented lagrangian* function, including the original objective $f^0(\mathbf{x}^N) + \mathbf{c}^{\top}\mathbf{x}$, and a term involving Lagrange multiplier estimates $\boldsymbol{\lambda}_k$. The modified penalty term $\frac{1}{2}\rho(\mathbf{f} - \tilde{\mathbf{f}})^{\top}(\mathbf{f} - \tilde{\mathbf{f}})$ is adjoined to enhance convergence from points far removed from a local minimum (Sargent and Murtagh (1973)). The term is dropped when it is evident that a minimum is approached (as measured by the error in satisfaction of Eq. (6-43) and relative change in estimates $\boldsymbol{\lambda}_k$).

The Lagrange multiplier estimates are chosen by using the values from the solution of the previous linearized problem (Robinson, 1972); these are easily obtainable as the first m_1 components of $\boldsymbol{\pi}$ at the solution.

In using MINOS to solve Eqs. (6-48)–(6-51), we use a quasi-Newton approximation to the reduced Hessian of Eq. (6-48). The cycle of major iterations can be described as follows:

Step 1 Set $k = 0$, and set $\boldsymbol{\lambda}_0$, $\mathbf{x}_0^N$ equal to some initial estimate.

Step 2 Having $\mathbf{x}_k^N$, $\boldsymbol{\lambda}_k$, and ρ, solve the linearly constrained problem, Eqs. (6-48)–(6-51), to obtain new quantities $\mathbf{x}_{k+1}$ and $\mathbf{g}(\mathbf{x}_{k+1})$. Partition $\mathbf{g}(\mathbf{x}_{k+1}) = (\mathbf{g}_B \;\; \mathbf{g}_S \;\; \mathbf{g}_N)$ to match the final basis and solve $\mathbf{B}^{\top}\boldsymbol{\pi} = \mathbf{g}_B$.

Step 3 Test $\mathbf{x}_{k+1}$ for convergence. If optimal, EXIT.

Step 4 Relinearize the constraints at $\mathbf{x}_{k+1}$. Order the columns of the new constraint matrix into $[\mathbf{B} \;\; \mathbf{S} \;\; \mathbf{N}]$ in the same sequence as before:

If $\|\mathbf{f}(\mathbf{x}_{k+1}^N) + A_1\mathbf{x}_{k+1}^L - \mathbf{b}_1\|/\|(1 + \mathbf{x}_{k+1})\|$ and $\|\boldsymbol{\lambda}_{k+1} - \boldsymbol{\lambda}_k\|/\|(1 + \boldsymbol{\lambda}_{k+1})\| \leqslant \varepsilon_c$, then set $\rho = 0$. Set $\boldsymbol{\lambda}_{k+1}$ = the first m_1 components of $\boldsymbol{\pi}$. Set $k = k + 1$ and repeat from Step 2. ε_c is some small convergence tolerance ($\sim 10^{-2}$).

This extension has been implemented (Murtagh and Saunders, 1980), and numerical experience confirms that the use of the lagrangian terms produces a generally faster and more reliable algorithm than simple sequential linearization.

6-4 INTEGER PROGRAMMING

Integer programming is a large subject worthy of a textbook in its own right, and indeed many excellent texts on the subject exist (see, for example, Garfinkel and Nemhauser, 1972). We shall consider only one technique for integer programming, the *branch and bound method*, since this is directly related to the simplex method. It is also the most generally successful technique, in spite of its rather pedestrian theoretical motivation, and it is the method used in nearly all commercial codes.

6-4-1 The Branch and Bound Method

Consider the following problem:

Maximize
$$\mathbf{P} = \mathbf{c}^{\mathrm{T}}\mathbf{x} \tag{6-52}$$
subject to
$$\mathbf{A}\mathbf{x} = \mathbf{b} \tag{6-53}$$
$$\mathbf{l} \leqslant \mathbf{x} \leqslant \mathbf{u} \tag{6-54}$$
$$x_j \text{ integer}, j = 1, \ldots, r$$

When $r < n$, the problem is called a *mixed-integer* program. When $r = n$, the problem is called a *pure-integer* program. The method is based on the observation that an optimal solution to Eqs. (6-52)–(6-54) will also satisfy

either
$$x_j \geqslant I_j + 1 \tag{6-55}$$
or
$$x_j \leqslant I_j \qquad j = 1, \ldots, r \tag{6-56}$$

where I_j is any integer between l_j and u_j. The idea is to *branch* – split the problem into two further problems:

one with bounds
$$l_j \leqslant x_j \leqslant I_j \tag{6-57}$$
the other with bounds
$$I_j + 1 \leqslant x_j \leqslant u_j \tag{6-58}$$

for a particular variable, j, $1 \leqslant j \leqslant r$, and solve each as a continuous linear-programming problem, ignoring the integer requirements. This process is repeated for different variables j and different integers I_j, and it will be evident that the effectiveness of such an approach is highly dependent on having a good systematic way of choosing j and I_j.

The choice of I_j can be resolved by solving Eqs. (6-52)–(6-54) as a continuous LP and setting

$$I_j = [x_j] \qquad \text{for some } j,\ 1 \leqslant j \leqslant r \tag{6-59}$$

where

$$x_j = [x_j] + f_j \qquad 0 \leqslant f_j < 1 \tag{6-60}$$

$[x_j]$ is the integer component of x_j, the (continuous) solution to the LP, and f_j is the fractional component. Performing the branching process, Eqs. (6-55) and (6-56), gives us two new problems in which Eqs. (6-57) and (6-58) respectively are added to the existing constraints.

We must now choose which of these two new problems to solve as a continuous LP, and as this procedure is repeated there gradually builds up a *master list* of problems remaining to be solved. Solution of each problem from the master list will result in two new problems, obtained by branching on one variable, $1 \leqslant j \leqslant r$, which has a noninteger value at the solution. The sequence of problems develops a tree-like structure, as shown in Fig. 6-3. After branching on a variable, the constraints in the two new problems corresponding to the new bounds are additional to those already existing for the "parent" problem. The tree structure shown in Fig. 6-3 is not necessarily complete; the branching of each parent problem terminates when one of three criteria is satisfied:

1. The solution is infeasible. This is increasingly likely to happen as the branching process continues, as more and more constraints of the form of Eqs. (6-57) or (6-58) are added to the existing set.

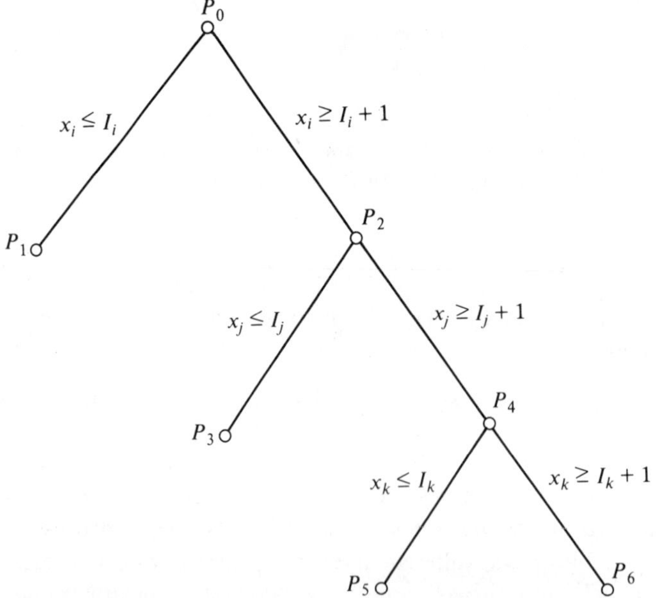

Figure 6-3 Tree structure of branch-and-bound algorithm.

2. The solution is less than the current best-known, integer-feasible solution. Adding further constraints by branching cannot possibly improve the value of the objective, so there is no point in continuing the branching process from this solution.
3. The solution is integer-feasible, i.e., $x_j, j = 1, \ldots, r$ has integer values. As more integer-valued upper and lower bounds are added to the variables with each new problem, the continuous solution will eventually have integer values (providing an integer-feasible solution exists).

The steps of the procedure are as follows:

Assume at iteration t we have the following information:

(*a*) A current best-known, integer-feasible solution x_0^t. This provides a lower bound for the optimal integer-feasible solution.
(*b*) A master list of problems remaining to be solved.
(*c*) A current problem selected to be the next one solved.

Step 1 *LP*. Solve the current problem as a continuous LP.

(*a*) If it is not feasible; or
(*b*) If it has an objective value x_0 less than x_0^t;

set $x_0^{t+1} = x_0^t$, $t = t + 1$, and go to Step 4.

Step 2 *BOUND*. If the solution is integer-feasible, set x_0^{t+1} equal to the objective value, $t = t + 1$, and go to Step 4.

Step 3 *BRANCH*

(*a*) Select *one* variable x_j, $1 \leqslant j \leqslant r$, which does not have an integer value, and set $I_j = [x_j]$.
(*b*) Form two new problems:

 (i) one with the lower bound on x_j replaced by $I_j + 1$;
 (ii) the other with the upper bound on x_j replaced by I_j.

(*c*) Select one problem to be solved, and add the other to the master list. Set $x_0^{t+1} = x_0$, $t = t + 1$.

Return to Step 1.

Step 4 *BACKTRACK*. Select a problem from the master list. Return to Step 1.

Evidently, the procedure could be altered to have both problems added to the master list in Step 3, followed by a move to Step 4. The rate of convergence now becomes dependent on the choice of variable j to branch on at each step, and the method of selecting problems from the master list. This choice can be aided by the calculation of *penalties*, proposed by Driebeek (1966) and developed further by Tomlin (1970).

Consider the pth basic variable in the optimal tableau of a parent problem,

$$(\mathbf{x}_B)_p = \beta_p - \sum_{j=1}^{n-m} \alpha_{pj}(\mathbf{x}_N)_j \tag{6-61}$$

and

$$\beta_p = [\beta_p] + f_p \qquad 0 \leqslant f_p < 1 \tag{6-62}$$

Without loss of generality, we may assume that the nonbasic variables $(\mathbf{x}_N)_j$ are zero. If $(\mathbf{x}_B)_p$ is the variable chosen to branch on, then imposition of a lower bound $[\beta_p] + 1$ must cause it to increase by an amount $(1 - f_p)$. The minimum amount that the objective function must decrease by is given by the "up" penalty

$$P_U = \min_{j|\alpha_{pj}<0} \left\{ (1 - f_p)\left(\frac{d_j}{-\alpha_{pj}}\right) \right\} \tag{6-63}$$

where d_j is the reduced cost of the jth nonbasic variable.

Similarly, imposition of the upper bound $[\beta_p]$ for the other derived problem must decrease the variable by an amount f_p. The minimum amount that the objective function must decrease by is given by the "down" penalty

$$P_D = \min_{j|\alpha_{pj}>0} \left\{ f_p\left(\frac{d_j}{\alpha_{pj}}\right) \right\} \tag{6-64}$$

For any nonbasic integer variables, we can replace the expression inside the brackets of Eqs. (6-63) and (6-64) by d_j, if it has a numerical value greater than that of the existing expression, since such an entering nonbasic variable must enter by at least a unit amount in order to remain integer-feasible.

Having calculated the penalties P_U and P_D, we may discontinue the branching process if

$$\max\{P - P_U, P - P_D\} \leqslant x_0^t \tag{6-65}$$

for all unsatisfied integer basic variables, where P is the objective value of the parent problem and x_0^t is the current best-known, integer-feasible solution. Otherwise, we branch on the basic variable p which corresponds to the largest penalty, and solve the opposite derived problem. This means that the problem with the largest penalty gets added to the master list in the hope of avoiding subsequent "backtracking" up the branches of the tree when an integer-feasible solution is found. In practice the procedure is often terminated once such a solution is found if it is within an acceptably small percentage of the original continuous solution.

When a branch is terminated, for whatever reason, the *best-projection* method (Forrest, Hirst, and Tomlin, 1974) can be used to select the next parent problem to branch on. This method predicts which parent problem will lead to the best integer solution. The prediction is based on estimating the decrease in objective value required to reduce the sum of integer infeasibilities to zero.

Consider the tree diagram in Fig. 6-3. At any iteration, there are a number of parent problems and we may suppose that by continuing to branch to the right as shown, an integer-feasible solution was eventually found. We must now choose which parent problem to backtrack to. An estimate of the decrease in objective value per unit decrease in the sum of fractions is given by

$$V_0 = \frac{P_0 - x_0^t}{F_0} \tag{6-66}$$

where P_0 is the objective value at the solution to the original continuous problem, F_0 is the sum of integer infeasibilities, min $\{f_p, 1 - f_p\}$, of the corresponding integer-infeasible basic variables, and x_0^t is the current best-known, integer-feasible solution. Thus if we have a number of parent problems, k, with objective value P_k and sum of fractions F_k, we choose the one with the highest value of $P_k - F_k V_0$. In place of x_0^t in Eq. (6-66), the user may supply an estimate of the optimal integer-feasible objective value.

Most commercial codes also allow the user to specify a priority list for the branching order, since knowledge of the physical nature of the problem and which variables are more important than others is often superior to any mechanical selection process.

The maximum objective value of all the problems remaining in the master list provides an upper bound V_1 on the optimal integer feasible solution. Again, commercial codes allow one to terminate when

$$\frac{V_1 - x_0^t}{V_1} \leqslant 0.1 \text{ (say)}$$

for then we are assured that we are within 10 percent of the optimal solution (which may take much longer to find).

6-4-2 Formulation of Integer Programming Problems

Integer programming problems are not easy to solve, but the speed of solution can be enhanced considerably by careful formulation. Williams (1978) illustrates how the same problem can be formulated in a number of different ways to give a faster solution rate, and the effort spent in prior analysis can result in enormous savings in computer time.

In contrast to the continuous linear-programming problem, it is generally advantageous to have the model as severely constrained as possible, and have the bounds on the variables as tight as possible, since this reduces the solution space to be examined. For example, a constraint of the form

$$13x_1 + 50x_2 + 11x_3 \geqslant 10$$

where x_1, x_2, and x_3 are all integers, can be replaced by the more severe constraint

$$x_1 + x_2 + x_3 \geqslant 1$$

Similarly, the constraint

$$13x_1 + 5x_2 + 11x_3 \geqslant 10$$

can be replaced by

$$x_1 + x_3 \geqslant 1$$

if x_2 is a *binary* (0, 1) variable, since even if x_2 takes the value 1 either x_1 or x_3 must also be 1 or more.

Binary variables can be used to model logical relationships. A logical statement T_j, which can be either TRUE or FALSE, can be modeled by a binary variable x_j which has a value 1 if T_j is TRUE and a value of 0 if T_j is FALSE. NOT T_j can be modeled as $(1 - x_j)$. Thus, for example, we can model:

1. T_1 or T_2 or T_3 is TRUE, as

$$x_1 + x_2 + x_3 \geqslant 1$$

2. One and only one of T_1, T_2, and T_3 is TRUE, as

$$x_1 + x_2 + x_3 = 1$$

3. T_4 is TRUE if and only if T_1 and T_2 and T_3 are all TRUE, as

$$0 \leqslant x_1 + x_2 + x_3 - 3x_4 \leqslant 2$$

4. T_4 is TRUE if and only if T_1 or T_2 or T_3 is TRUE, as

$$-2 \leqslant x_1 + x_2 + x_3 - 3x_4 \leqslant 0$$

The use of binary variables to model set-up costs is a well-established application. Consider a situation in which we have the cost expressed as

$$\text{Cost} = \begin{cases} 0 & \text{if } x_2 = 0, \\ c_1 + c_2 x_2 & \text{if } x_2 > 0. \end{cases}$$

c_1 is the set-up cost associated with x_2 attaining a nonzero value. This, for example, would be the cost of setting up the tooling, jigs, and fixtures in a manufacturing process.

We introduce a binary variable x_1 with the property

$$\begin{aligned} x_1 &= 0 \qquad \text{if } x_2 = 0 \\ x_1 &= 1 \qquad \text{if } x_2 > 0. \end{aligned}$$

We may now model the cost as

$$c_1 x_1 + c_2 x_2$$

and force the required property on x_1 by the constraint

$$x_2 - mx_1 \leqslant 0$$

where m is a number larger than any value x_2 will take. (If $x_2 = 0$, the cost term $c_1 x_1$ will force x_1 to be zero also.)

Another similar application is the *minimum batch size* problem, where

either $$x_2 = 0$$

or $$x_2 \geqslant l$$

l is the minimum batch size. This can be modeled with a binary variable x_1 and two constraints,

$$x_2 - lx_1 \geqslant 0$$
$$x_2 - mx_1 \leqslant 0$$

The two constraints force $x_2 = 0$ if $x_1 = 0$, while the first constraint forces $x_2 \geqslant l$ (and the second constraint is inactive) if $x_1 = 1$. Thus, x_1 (0, 1) represents the (NO, YES) decision to proceed with the batch.

6-4-3 Special Ordered Sets

This is a particularly useful concept due to Beale and Tomlin (1970). There are two types of special ordered sets:

Type 1 – only one variable in the set may be nonzero.
Type 2 – only two variables in the set may be nonzero, and they must be adjacent.

We have already seen an instance of Type 2 sets in Sec. 6-1 on separable programming, but the concept is exploited much more powerfully than that. Although the variables in the sets are not required to be integer, they are handled using the same branch-and-bound approach as in integer programming.

Type 1 sets are used for multiple-choice problems, such as facilities location or personnel assignment, where for example there is a constraint of the form

$$\lambda_1 + \lambda_2 + \cdots + \lambda_n = 1 \tag{6-67}$$

Since only one variable is allowed to be nonzero, it takes the value of unity.

However, there need not necessarily be constraints of the form of Eq. (6-67), and in such cases a noninteger value may be meaningful. For example, λ_k may represent the capacity of one of a set of mutually exclusive facilities.

Type 2 sets can be used to find global solutions to nonconvex problems. Using the rules of separable programming as described in Sec. 6-1, we can only obtain a local solution to such problems. However, using the branch-and-bound approach we may assign two "markers" to each Type 2 set, so that the only nonzero members must be between the markers. We then branch on a set by choosing a member λ_k and defining two new problems: one with the left-hand marker between λ_{k-1} and λ_k, so that all members before λ_k must be zero; the other with the right-hand marker between λ_k and λ_{k+1}, so that all members after λ_k must be zero. This enables us to divide the nonconvex problem into a series of piecewise convex problems and evaluate the optimum of each.

Type 1 sets can be handled in the same manner, except we branch on a set by placing both markers between λ_k and λ_{k+1}: one problem has all members before λ_{k+1} zero, and the other problem has all members after λ_k zero. One or other of these conditions must be true at the final solution, so it is analogous to branching on an integer variable, with the left-hand and right-hand markers performing the same role as the lower and upper bounds on the integer variable.

PART TWO

APPLICATIONS

CHAPTER

SEVEN

PROBLEM FORMULATION

7-1 INTRODUCTION

This chapter, indeed this whole of Part II of this book, is intended to serve as a guide on how to go about formulating and solving large real-world problems using commercially available LP codes. It is a discussion of LP modeling rather than one on LP techniques.

In order to avoid the discussion degenerating into a series of general comments of no particular value in themselves, we will be using a specific problem to illustrate our discussion throughout the rest of the book. Thus, by construction we are removing some of the generality, but it is hoped that in return we gain something tangible which can be used as a basis for other applications.

The example chosen is a typical investment problem in the oil industry. It is of the type solved regularly by most large oil companies, and the numbers used reflect a real-life situation (making obvious allowances for inflation and changes in technology). While analogies can be made directly with other process industries, it is also true in the broader sense that most activities can be represented as a "flow" process in which material, manpower, and money interact to produce desired end results represented as "products."

7-2 DEFINING THE BOUNDARIES: SCOPE VERSUS DETAIL

Much of the philosophy of problem formulation can be summarized in two principles:

1. The boundaries of the problem should be as wide as possible. Otherwise, the model may be performing suboptimization and not grasping the real issues involved. The danger is that one section may be optimized at the expense of others, and to the detriment of the overall organization.
2. The model should be as simple as possible. The model must be validated, checked, and understood, and the issues addressed by the model must be transparent to both the modeler *and* the manager.

While in theory the two ideals should not conflict, in practice they often do, simply because of the human element involved in data collection, data input, error checking, and interpretation of output. These factors combine to limit the size of model which can be analyzed satisfactorily, and if we are to use model size as a limiting factor then an increase in scope must mean a decrease in detail.

We are thus led to the concept of a hierarchy of models, where scope increases and detail decreases as we proceed up the levels of the hierarchy. The higher levels in turn produce constraints and objectives for the lower levels. A typical scheme for an industrial or commercial organization is illustrated in Fig. 7-1.

The scope of a model includes its time boundaries; the planning horizon generally increases with level of hierarchy. Long-term policy models of an entire organization would contain little of the day-to-day operating details, yet a production planning model of a specific plant would consist mostly of such detail. An excellent account of the use of model hierarchies at ICI is given by Stephenson (1970).

An example of model hierarchy is a national energy model (Smith, Lucas, and Murtagh, 1976a), in which forecast future demands for energy and availability of resources are postulated, and the model determines the least-cost mix of production and distribution facilities and interfuel substitution to satisfy the specified end-use demand. A diagram illustrating the system under investigation is shown in Fig. 7-2.

Now each component or subsystem of this can in turn be modeled in greater detail, where the overall model has specified resource availabilities and output requirements for each of the subsystems. For example, an electricity submodel (Smith, Lucas, and Murtagh, 1976b) determines the generating plant required to

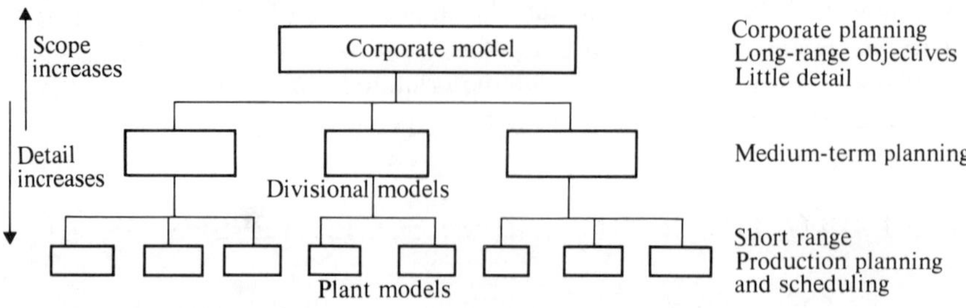

Figure 7-1 Model hierarchy for a large corporation.

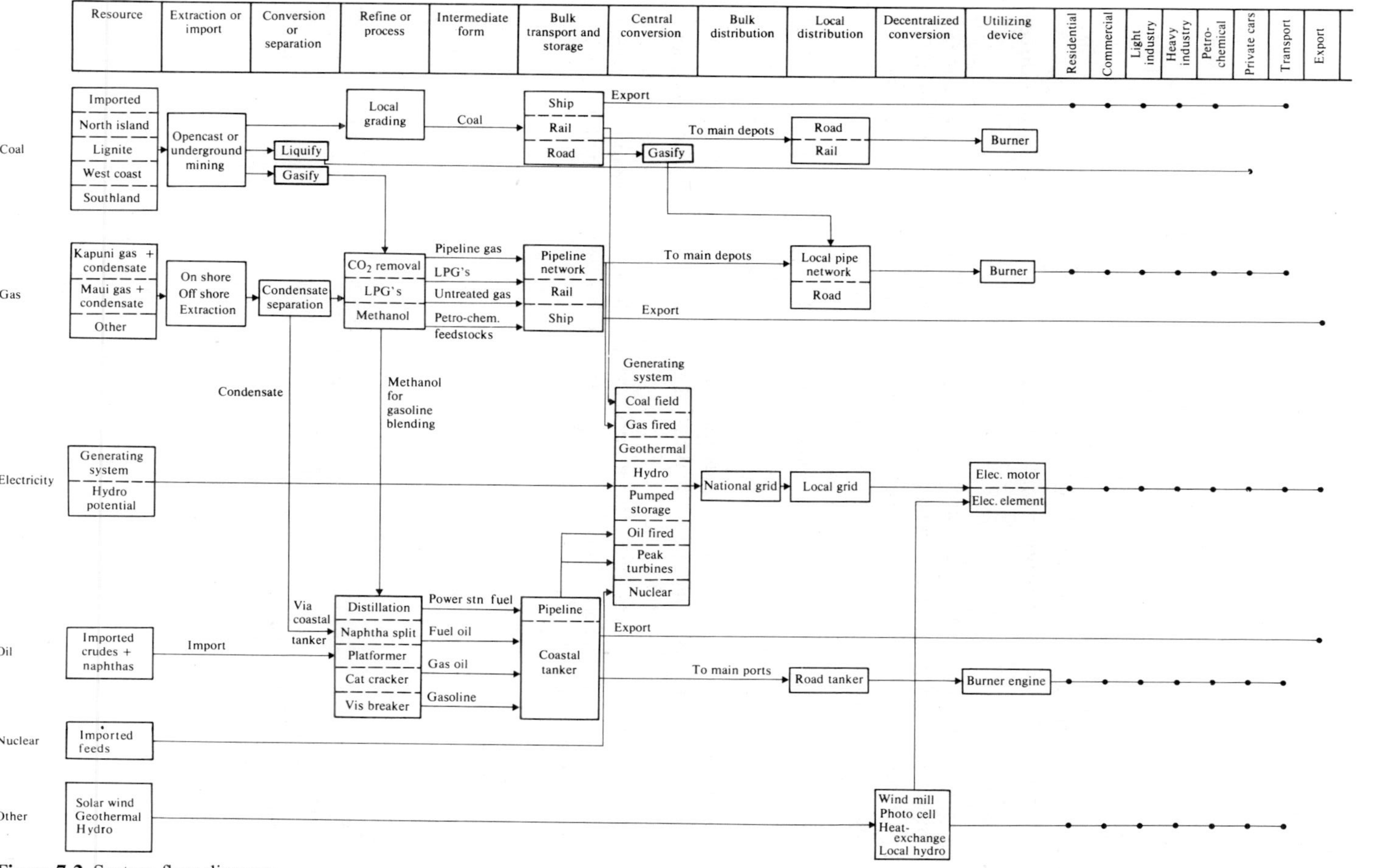

Figure 7-2 System flow diagram.

produce the specified demand for electricity. This submodel can now consider the load–duration characteristics of the demand, in order to determine the optimum configuration of generating plant. (Gas turbines, having low capital cost and high running cost, are favored for peak load, while coal-fired thermal or nuclear plant having high capital cost and low running cost are favored for base load.) A load–duration curve, as shown in Fig. 7-3, shows the number of hours a year the power demand (in megawatts) exceeds a given value. For LP modeling it is approximated by a piecewise-constant function (shown dotted in Fig. 7-3).

An oil refinery submodel can similarly be developed, but since we will be discussing a specific oil refinery model in detail, we will not pursue this here.

The boundaries of the problem are determined largely by the following three considerations:

1. *The terms of reference.* The objectives of the study, however loosely defined, influence in large measure what goes into the model. Engineers enjoy a particular advantage in this respect as the terms of reference are usually cut and dried, with the objective expressed in terms of maximum profit, minimum cost, or possibly minimum consumption of certain resources. However, the social scientist is usually given the task of maximizing "social benefit," and is left in the unenviable position of having to assign a "utility" to various socially desirable activities. The results of his study can often then, unfortunately, be questioned just on the basis of his definitions.
2. *Physical boundaries.* The spatial aspects of the study need detailed consideration. If there is more than one location where activities occur, we must model the associated physical distribution processes. These may include storage, warehousing, transportation, and the further problems of scheduling and facilities location.

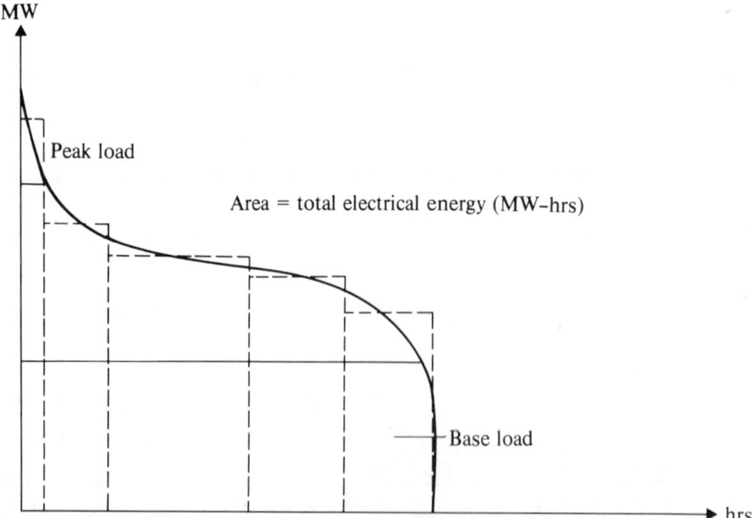

Figure 7-3 Load–duration curve for power generation.

3. *Time boundaries.* The temporal aspects of the study give rise to the greatest dilemma facing the analyst. There is usually a well-defined planning horizon and the choice is whether to model the system in time stages so that profiles in time are obtained, or to model *steady-state* behavior at a particular point in time.

   If the model is time-staged, the size of the model will tend to multiply as the number of time periods considered. Such models remain conceptually simple, so the real burden is placed on solvability within a reasonable length of time in the computer rather than on the ability to interpret the accompanying large volume of output. (Most of the really large models which occur in practice are time-staged.)

   It is often adequate to model the system at a particular point in time, say a "target year," and repeat the calculation at a number of time increments. The danger in this approach is that, since the problem is not solved for the whole planning horizon at once, such time profiles as the optimal depletion rate for scarce resources cannot be determined. However, the availability of resources may well be an uncertain estimate, and dependent on factors outside the scope of the model. The moral is therefore to examine carefully whether time-staged profiles of the variables in the model are really needed, or whether the same information can be gleaned by repeating a steady-state calculation for a number of different target years.

A cautionary note regarding detail is also worth making. One of the arguments in strong favor for having a hierarchy of models is that the level of detail is consistent in each. A trap quite easily fallen into is to have a model with wide scope and little detail except for one particular aspect of the system which is modeled in extremely fine detail. The danger here is that this will give rise to a number of rows which are almost (but not quite) identical compared with the rest of the A matrix. For a variety of reasons, such as frequent interchanges of relatively dense columns, fill-in, and rounding error, the solution time will slow considerably.

Example An oil refinery investment problem An oil refinery was built a number of years ago and the firm now wishes to expand capacity. We have been asked to conduct a study of several refinery expansion options to determine what combination of units and capacities give the maximum net profit.

A block flow diagram of the existing refinery is shown in Fig. 7-4. Crude oil enters the crude-vacuum unit, which separates the crude oil into a number of streams according to their boiling-point range. Some of the heavier liquids (atmospheric gas oil and vacuum gas oil) go to a hydrocracker (*unicracker*) which splits the molecules and adds hydrogen, thereby producing lighter liquids. The saturated gas unit separates the gases and light liquids into LPG (propane and butane), lighter gases for plant fuel gas, and liquids heavier than butane, some of which pass to the reformer (*platformer*) and others get blended into gasoline

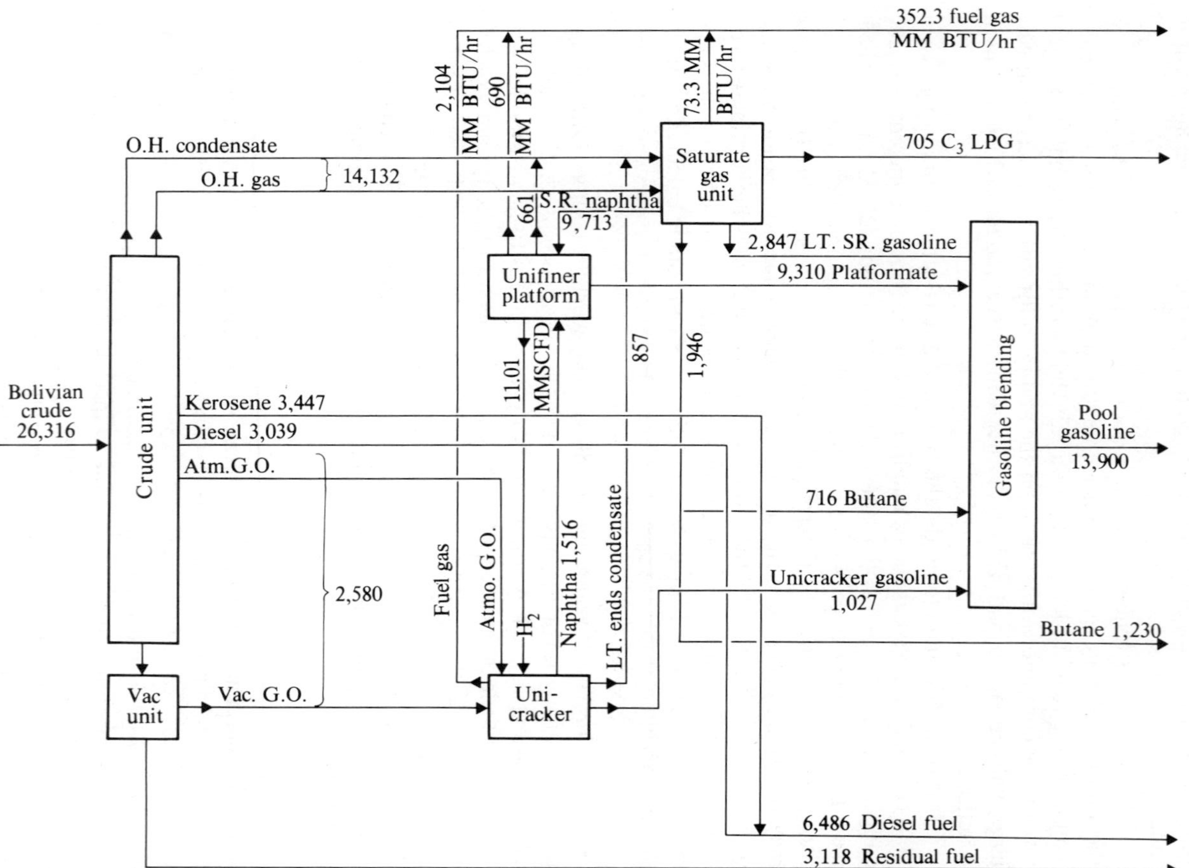

Figure 7-4 Block flow diagram of refinery (*All figures are based on BSD unless otherwise stated*).

directly. The reformer is needed to boost the octane number of these liquids; if there were no octane-number specification, all the liquids could ostensibly go directly into gasoline. We shall see later that octane number is only one of many quality specifications on the gasoline and these in turn give rise to many constraints. (The reader may be reassured that a knowledge of the workings of the process units is *not* required; nor is there any need for any prior knowledge of the process streams.)

The block flow diagram can be regarded as representing steady-state behavior at design capacity. Thus the basis for our model will be daily operation with flow quantities expressed in thousands of barrels per stream-day (MBSD), i.e., per operating day. (Note that in Fig. 7-4 the flows are expressed in barrels per stream-day, so should be divided by a thousand to obtain them in MBSD.)

The investment options to be considered are as follows:

1. The crude-vacuum unit overhead capacity can be increased from 14.132 to 23.26 MBSD by an investment of \$1,200,000 in additional pumps and heat exchangers.
2. The hydrocracker was constructed with 50% excess reactor capacity. Thus the feed capacity can be increased from 2.58 to 3.87 MBSD by adding peripheral equipment at a cost of \$4,000,000.
3. A new fluid catalytic cracking unit (FCCU) could be constructed. This performs the same cracking operation as the hydrocracker, but hydrogen is not added so different compounds are produced. It is desirable to construct an alkylation unit in conjunction with the catalytic cracker for further treatment of some of the compounds produced. A combined cost of \$35,400,000 is estimated for an FCCU and alkylation unit of capacity 7.260 MBSD of FCCU feed. We are to investigate other capacities for the FCCU, with investment cost varying according to a six-tenths power law to reflect the economies of scale:

$$\text{Cost} = 35.4\left(\frac{\text{FCCU capacity}}{7.26}\ (\text{MBSD})\right)^{0.6} \ \$\ \text{MM}$$

7-3 USE OF FLOWCHARTS

The physical nature of the system being modeled can invariably be represented by a flowchart. A very simple example is the block flow diagram described in the previous section (Fig. 7-4), although this in itself does not contain enough detail to be of much use to us. The properties of flowcharts are as follows:

1. Lines represent flow streams of *material*, characterized by certain attributes. They need not necessarily be physical material; for instance flow of information, or cash, could equally be represented in this way. If two streams of material have different attributes which are to be distinguished in the model, then we must draw separate lines for each.

2. Boxes represent units of plant and equipment, or more generally *subsystems*, each performing a specific function. The attributes of the flow streams change, so the units are a terminating point for one set of lines and a source point for another.
3. By convention, the direction of flow is from left to right. Thus in describing an industrial process, the raw material enters on the left of the flowchart, and the finished products flow into *demand* columns on the right. These demand columns are particularly useful when more than one process stream is blended into a final product, as we shall see in our example.

Example The flowchart for our problem is shown in Fig. 7-5. It is considerably more detailed than the block flow diagram of Fig. 7-4, particularly since every stream we wish to consider in our model is represented by a line.

Two types of crude oil feedstock are available: Bolivian crude and Columbian crude. Since they have different properties they are represented by two lines, labeled BOLCDF and COLCDF respectively.

The operation of the reformer is represented as three separate processes. The unifiner removes sulphur from the stream labeled HSRNPD; this requires hydrogen, so there is represented a consumption of the hydrogen stream NETH2P. (Note also the hydrocracker consumes NETH2P.) The reaction process inside the reformer is considered at two levels of severity: one which produces reformate labeled 95 REFP, and one which produces reformate labeled 90 REFP. Other products are 95 BBP and 90 BBP, as well as hydrogen, NETH2P. The FCCU and alkylation unit have also been added.

The demand columns on the right-hand side represent three grades of gasoline (premium, intermediate, and regular) as well as LPG (bottled gas), JP4 (jet fuel), diesel, and residual fuel oil. Note in particular the large number of process streams blended into gasoline. (The dots in the demand columns indicate that the stream enters the product blend.)

7-4 DESCRIPTIVE CONSTRAINTS

As their name implies, these constraints describe the operation of the system under investigation. They are essentially a group of material balance equations, where again "material" may refer to any attribute defined with consistent units, for example mass, energy, or cash.

The fact that the balance equations must be linear rules out representation of nonlinear interactions between variables such as complex chemical reactions, but first-order effects of changes in operating conditions can be modeled and this is usually adequate. We will discuss this point further in Sec. 8-6.

While balances could be carried out around an envelope encircling any convenient portion of the flowchart, for example each process unit, they are usually carried out around each stream represented by a line in the flowchart. Thus, material entering the stream (yielded from possibly more than one unit) is

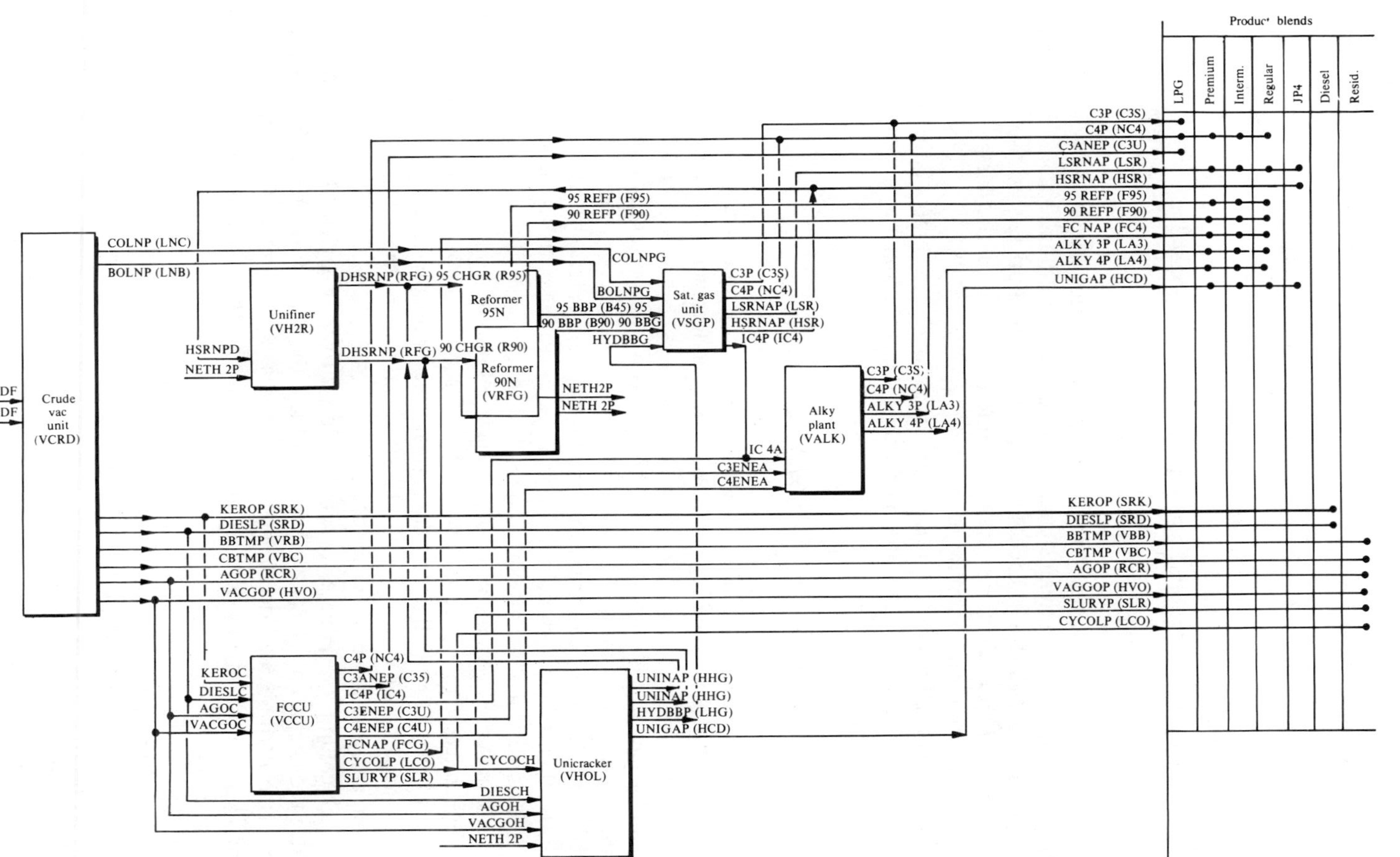

Figure 7-5 Refinery LP flowsheet.

equated to that leaving the stream (again as a feed to possibly more than one unit).

For steady-state operation, the equations can be expressed in the form

$$-\text{Input} + \text{output} = 0 \tag{7-1}$$

For time-staged operation, the equations can be expressed as

$$-\text{Input} + \text{output} + \text{accumulation} = 0 \tag{7-2}$$

where accumulation is the net gain during the time period.

Suppose there are K streams entering a particular unit and x_k, $k = 1, \ldots, K$, are the amounts (MBSD) in each stream entering the unit. Let us also suppose that within the unit each stream yields an amount a_{ik} of a particular product per unit feed. Thus the total amount yielded is $\sum_{k=1}^{K} a_{ik}x_k$ (MBSD).

Let us suppose further that this particular product itself enters only one unit, and x_j is the amount (MBSD) entering the unit. Thus the material balance row (i) for that stream is

$$-\sum_{k=1}^{K} a_{ik}x_k + 1x_j = 0 \tag{7-3}$$

Now since each stream consists of products from unit(s) and feed(s) to subsequent unit(s) or final product (see Fig. 7-5), we adopt the following convention:

1. Define a row, i, for each flow stream; this row will be the material balance equation for the stream.
2. Define a column for each unit that the flow stream enters, with a corresponding $+1.0$ entry for each column. This column corresponds to the variable x_j, whose numerical value gives the amount of that flow stream entering the unit. (Recall in Sec. 2-1 we mentioned that each column of A has a unique variable x_j associated with it; we can therefore refer to "columns" and "variables" synonymously.) Note that the flow stream may possibly enter more than one unit, and each $+1\ x_j$ term (cf. Eq. (7-3)) would then be the amount of the flow stream entering the unit.
3. In each column (variable x_k) where the flow stream was yielded as a product, enter the coefficient $-a_{ik}$. Note we can have the same flow stream (i.e., with the same attributes) yielded from more than one unit as well as from more than one feed stream, giving extra yield coefficients in the same material balance row.

Since all the entries in (2) and (3) above are made in the same row, the net result is a material balance equation of the form of Eq. (7-3) (with more than one $+1\ x_j$ term if the stream enters more than one unit).

To summarize: a material balance *row* corresponds to a *stream*; characterized by a particular set of attributes, but with possibly more than one entry point and exit point. A *column* x_j is defined at each new point the stream enters a *unit*.

A further general convention which is retained in all types of constraints where appropriate, is to have negative coefficients where material is yielded into the system and positive coefficients where material is absorbed from the system (see, for example, the utility balance constraints in Sec. 7-5).

Example As an example, we shall consider the streams BOLNP (Bolivian naphtha) and COLNP (Columbian naphtha) yielded from the crude unit, and the stream HSRNAP (heavy straight run naphtha) yielded from the saturated gas unit (see Fig. 7-5).

Bolivian crude (BOLCDF) yields 0.5370 bbl BOLNP per bbl BOLCDF,† and Columbian crude (COLCDF) yields 0.2931 bbl COLNP per bbl COLCDF. These two streams enter the saturated gas unit and yield different proportions of the product streams C3P, C4P, IC4P, LSRNAP, HSRNAP. The HSRNAP stream then enters both the unifiner and JP4 product blend.

The matrix representation of the material balance equations for these streams is as follows:

	Crude unit VCRD		Sat. gas unit VSGP		Unifiner VH2R	JP4 blend BJP4	
	BOL	COL	LNB	LNC	HSR	HSR	
LNC		−0.2931 . .		1.0 . .		. .	= 0
LNB	−0.5370		1.0				= 0
C3P			−0.0277	−0.0112			(7-4)
NC4			−0.0563	−0.0378			
IC4			−0.0170	−0.0099			
LSR			−0.1990	−0.1502			
HSR			−0.6873	−0.7953	+1.0	+1.0	= 0

Thus, considering just these streams, the three material balance equations would be

$$-.2931\, x_{\text{VCRDCOL}} + 1.0\, x_{\text{VSGPLNC}} = 0 \qquad \text{(Row MVOLLNC)}$$
$$-.5370\, x_{\text{VCRDBOL}} + 1.0\, x_{\text{VSGPLNB}} = 0 \qquad \text{(Row MVOLLNB)} \qquad (7\text{-}5)$$
$$-.6873\, x_{\text{VSGPLNB}} - .7953\, x_{\text{VSGPLNC}} + 1.0\, x_{\text{VH2RHSR}} + 1.0\, x_{\text{BJP4HSR}} = 0 \qquad \text{(Row MVOLHSR)}$$

Note that rather than use i's and j's, we use descriptive labels. (These labels are taken partially from the stream names and partially from the unit names; this will be discussed in Chapter 8.) The other yield streams from the saturated gas unit (C3P, C4P, IC4P, LSRNAP) also in turn become feed streams to other units or final products. If we continue using the same convention we will eventually have a material balance row for every stream in the flowchart. (The prefix MVOL is used to distinguish these as material balance rows.) The streams C3P and C4P are also yielded from the FCCU and alkylation unit (see Fig. 7-5). Thus these will have extra yield coefficients in their respective material balance rows.

† bbl = barrel.

Note also that we wished to distinguish between naphtha from Bolivian and Columbian crude as they had different attributes; in particular, they gave different proportions of product yields in the saturated gas unit. Thus they each had separate lines in the flowchart and, correspondingly, separate material balances. However, in each of the product yields from the saturated gas unit (e.g., HSRNAP) we make no distinction between their attributes, so we have just the one material balance row for each.

An observation worth making at this point is that in the layout of the matrix as shown in Eq. 7-4, the columns tend to be grouped together corresponding to each unit in the flowchart. We shall complete all the material balance equations in Chapter 8.

7-5 RESOURCE AVAILABILITY/PRODUCT DEMAND CONSTRAINTS

These constraints can be dealt with readily. In their simplest form the resource availability constraints are upper bounds on variables representing consumption, and product demands are lower bounds on variables representing production. It is possible to express the resource availability constraints in the following form:

$$a_{i1}x_1 + a_{i2}x_2 + \cdots + a_{ij}x_j + \cdots + a_{in}x_n \leqslant b_i \tag{7-6}$$

where a_{ij} is the consumption of the ith resource per unit of the activity x_j, $j = 1, \ldots, n$, and b_i is the total resource available.

However, if we defined a new variable, say x_{n+1}, which represented total consumption, then the constraint would be of the form:

$$a_{i1}x_1 + a_{i2}x_2 + \cdots + a_{ij}x_j + \cdots + a_{in}x_n - x_{n+1} = 0 \tag{7-7}$$

$$x_{n+1} \leqslant b_i \tag{7-8}$$

It would appear that we are adding another row, but recall from Sec. 5-1 that Eq. (7-8) can be handled using simple upper bounding.

By defining a_{ij} as the production of the ith product per unit of activity x_j, $j = 1, \ldots, n$, and reversing the sign of the inequality, we get analogous equations for the product demand constraints, with b_i representing total demand. Note that the capacity of plant and equipment can also be expressed as a resource availability.

It is possible to express price–volume relationships in both resource availability and product demand. These will be discussed in Sec. 7-7.

Example We will consider the following four types of constraint:

1. Availability of crude oil.
2. Utility requirements (steam, electricity, cooling water, process heat, and hydrogen).

3. Capacity of units.
4. Product sales.

1. Availability of crude oil
Supplies of Bolivian crude are limited to 26,316 BSD.
Supplies of Columbian crude are limited to 21,052 BSD.
If flow quantities are expressed in thousands of barrels per stream day (MBSD), we have

$$x_{\text{VCRDBOL}} \leq 26.316 \qquad \text{(Row MVOLBOL)}$$
$$x_{\text{VCRDCOL}} \leq 21.052 \qquad \text{(Row MVOLCOL)}$$

2. Utility requirements The utility requirements for each unit are based on the feedstreams entering that unit. Thus we can have a row of the form of Eq. (7-7), where a_{ij} represents the requirement of utility i per unit of activity x_j, $j = 1, \ldots, n$.

As an example, let us consider the electric power requirements. We have a row, labeled UBALKWH, with a nonzero a_{ij} term for every feedstream in the flowchart, representing the kW-hr per BSD of feed:

$$0.302\, x_{\text{VCRDBOL}} + 0.384\, x_{\text{VCRDCOL}} + 0.495\, x_{\text{VSGPLNB}} + 0.721\, x_{\text{VSGPLNC}} + 0.495\, x_{\text{VSGPLHG}} + \cdots - x_{\text{PURCKWH}} = 0 \qquad \text{(Row UBALKWH)}$$

This is of the form of Eq. (7-7), and the value of x_{PURCKWH} is the sum of all the individual power requirements.

The same type of equation is repeated for all utilities, so that each feedstream has an associated list of utility requirements. Thus, for example, the columns representing x_{BOLCDF}, x_{COLCDF}, and x_{BOLNP} have the following entries in the utility rows:

	VCRD			VSGP		UTIL			PURC		
	BOL	COL		LNB		HYL	STM	FUL	H20	KWH	
MSCFHYL	0	0	...	0	...	−1.0					
UBALSTM	.003	.003	...	.0130	...		−1.0				= 0
UBALFUL	.0587	.1053	...	.0506	...			−1.0			= 0
UBALH20	.15	.185	...	.209	...				−1.0		= 0
UBALKWH	.302	.384	...	.495	...					−1.0	= 0

The MSCFHYL row will have positive coefficients where hydrogen is required (in the hydrocracker and unifiner) and negative coefficients where it is produced (in the reformer). We shall complete all the entries of this matrix in Chapter 8. (Note that the prefix UBAL indicates a utility balance row, except MSCF, which indicates the units of the row: MSCF per day.)

*3. **Capacity of units*** the sum of the feedstreams entering a unit must be no greater than the feed capacity of the unit. Thus we have, from Fig. 7-5: For the hydrocracker:

$$x_{\text{VHOLLCO}} + x_{\text{VHOLSRD}} + x_{\text{VHOLRCR}} + x_{\text{VHOLHVO}} \leqslant 3.87 \qquad \text{(in units of MBSD)}$$

For the FCCU:

$$x_{\text{VCCUSRK}} + x_{\text{VCCUSRD}} + x_{\text{VCCURGR}} + x_{\text{VCCUHVO}} \leqslant 7.26 \qquad \text{(the RHS is to be varied parametrically)}$$

Sometimes the capacity is expressed as a restriction on the products leaving the unit, or as certain limitations on specific streams. In our example, the sum of the crude unit overhead streams (BOLNP and COLNP in Fig. 7-5) must be no greater than 23.26 MBSD:

$$x_{\text{VSGPLNB}} + x_{\text{VSGPLNC}} \leqslant 23.26$$

Also, the stream VACGOP from the crude-vacuum unit is limited to 5.25 MBSD.

It follows from our discussion in Sec. 7-4 and the flows in Fig. 7-5 that in the material balance row for this stream (row label MVOLHVO) there will be +1.0 entries in the columns labeled VCCUHVO and VHOLHVO, corresponding to it entering as a feedstream to the FCCU and hydrocracker respectively, as well as in the column BRSDHVO, where it enters the residual fuel oil blend. The capacity constraint is thus expressed as

$$x_{\text{VCCUHVO}} + x_{\text{VHOLHVO}} + x_{\text{BRSDHVO}} \leqslant 5.25$$

The capacity of the alkylation unit expressed as the sum of the streams ALKY3P and ALKY4P (in Fig. 7-5) is limited to 10.0 MBSD:

$$x_{\text{VALKLA3}} + x_{\text{VALKLA4}} \leqslant 10.0$$

Finally, the reformer reactor capacity must be considered. As mentioned in Sec. 7-3, the reformer reactor is operated at two levels of severity: one producing a 90 RON (Research Octane Number) reformate and one producing a 95 RON reformate. It is run under so-called *blocked operation*, which means it is run for a period at one severity, and then run at the other severity for a similar period. With this mode of operation, two distinct products are produced; thus as far as the LP model formulation is concerned, there are actually two reformer reactors, one at 95 RON and one at 90 RON. However, the capacity constraint is expressed in annual average operation.

At 95 RON, the capacity is 10.35 MBSD of feed 95 CHGR (Fig. 7-5). At 90 RON, the capacity is 13.455 MBSD of feed 90 CHGR. Then by linear interpolation,

$$x_{\text{VRFGR90}} + 1.3\, x_{\text{VRFGR95}} \leqslant 13.455$$

4. Product sales We will assume that there is a market for all the products produced. However, the three grades of gasoline must conform to the following constraints:

$$\text{Percent premium} \leqslant 36$$
$$\text{Percent regular} \geqslant 35$$

If the three grades of gasoline product are represented by x_{SELLPRE}, x_{SELLINT}, and x_{SELLREG} respectively, then these constraints can be expressed as:

$$0.64\, x_{\text{SELLPRE}} - 0.36\, x_{\text{SELLINT}} - 0.36\, x_{\text{SELLREG}} \leqslant 0$$
$$0.35\, x_{\text{SELLPRE}} + 0.35\, x_{\text{SELLINT}} - 0.65\, x_{\text{SELLREG}} \leqslant 0$$

7-6 EXTERNALLY IMPOSED SPECIFICATIONS

A number of constraints on the system could be regarded as externally imposed. Product quality specifications are usually imposed by statutory authorities and are subject to legislation. Similarly, environmental considerations impose limitations on the products (for example, the amount of sulphur in fuel oil) and on the operation of plant and equipment (for example, the quality of effluent wastewater, which would manifest itself as an extra cost).

Consider the situation where a number of different flow streams are blended to form a final product. If p_i is a property associated with the ith blend stream and p_b is the minimum property required for the blended product, then we have

$$p_1 x_1 + \cdots + p_i x_i \geqslant p_b x_b \tag{7-9}$$

where the left-hand side is summed over all flow streams entering the blend and x_b is the amount of blended product produced. The sign of the inequality is obviously reversed for a maximum property of the blended product; in our example above, the maximum percentage sulphur in fuel oil would be specified.

It is more convenient to rearrange Eq. (7-9) to the form

$$-p_1 x_1 - \cdots - p_i x_i + p_b x_b \leqslant 0 \tag{7-10}$$

and it will be seen that the pattern of Eq. (7-10) is similar to that of the material balance equations (7-3).

Example The product quality specifications listed in Table 7-1 are considered adequate for the purpose of investigating investment options. (Many refinery models, particularly those used for production planning in existing refineries, have large numbers of quality specifications.)

Table 7-1 Product quality specifications for refinery example

Product	Specifications
Premium gasoline	RON $\geqslant$ 100.3 RVP $\leqslant$ 9.5 psi 200 °F $\leqslant$ ASTM 50% $\leqslant$ 230 °F
Intermediate gasoline	RON $\geqslant$ 96.5 RVP $\leqslant$ 9.5 psi 200 °F $\leqslant$ ASTM 50% $\leqslant$ 230 °F
Regular gasoline	RON $\geqslant$ 90.5 RVP $\leqslant$ 9.5 psi 200 °F $\leqslant$ ASTM 50% $\leqslant$ 230 °F
LPG	None – any combination of propane-butane
JP4	2 psi $\leqslant$ RVP $\leqslant$ 3 psi
Diesel	None – straight run diesel plus kerosene
Fuel oil	SFS $\leqslant$ 175 at 122 °F

The terms RON (Research Octane Number), RVP (Reid Vapor Pressure), ASTM 50% (the temperature at which 50% of the mixture is vaporized), and SFS (a viscosity constraint), need not concern the reader; they are simply properties p_b required of the blended product, and the properties p_i of the blend streams are known constants.

As an example, consider the viscosity constraint on fuel oil. The specification on the blended product corresponds to a linear blending number of 10.1. The linear blending numbers of the blend streams are as follows:

Stream names (Fig. 7-5)	Linear blending nos.	Column label
BBTMP	10.1	BRSDVBB
CBTMP	12.63	BRSDVBC
VACGOP	8.05	BRSDHVO
AGOP	6.90	BRSDRCR
SLURYP	8.05	BRSDSLR
CYCOLP	4.40	BRSDLCO

Thus the constraint is

$$10.1\, x_{\mathrm{BRSDVBB}} + 12.63\, x_{\mathrm{BRSDVBC}} + 8.05\, x_{\mathrm{BRSDHVO}} + 6.90\, x_{\mathrm{BRSDRCR}} + 8.05\, x_{\mathrm{BRSDSLR}} + 4.40\, x_{\mathrm{BRSDLCO}} - 10.1\, x_{\mathrm{BELLRSD}} \leqslant 0$$

Note that it is a maximum property specification on the blended product, so the sign of the coefficients is the reverse of those in Eq. (7-10).

The product quality constraints are best dealt with using tables; such tables for all the specifications in our example problem will be presented in Sec. 8-2.

7-7 DEVELOPMENT OF OBJECTIVE FUNCTION

The objective function in the model usually consists of some or all of the following components:

1. Value of the product.
2. Capital cost of plant and equipment.
3. Cost of resources.
4. Operating and maintenance costs.

1. Value of the product If the system being modeled is a commercial undertaking, then the value of the product is described in terms of dollars and cents. If not, then the objective will be to maximize some form of social benefit, and the value of the output from the system must be described in terms of *utility*; differences in how this is defined may well give rise to different answers. For example, in hospital planning, an objective described in terms of maximizing patient throughput may not give the best service to the community. Similarly, in scheduling a computer system, maximizing job throughput will give priority to small jobs; this may not be an effective yardstick to measure the system's performance.

The application is straightforward: if x_i is the quantity of "product" and c_i is the value per unit, then we get a $c_i x_i$ term in the objective function. However, it may be necessary to go further than this: if we can sell so much at a certain price and more at a lower price, then we have a type of price–quantity curve. An example is the ore-processing plant discussed in Sec. 4-6, where the price–quantity curve is as shown in Fig. 7-6.

2. Capital cost of plant and equipment If the model is based on steady-state operation at a particular point of time, then all costs must be measured on a per day (or per year) basis. This is quite straightforward for all costs and values represented as a cost per unit of material flowing. However, in the case of capital

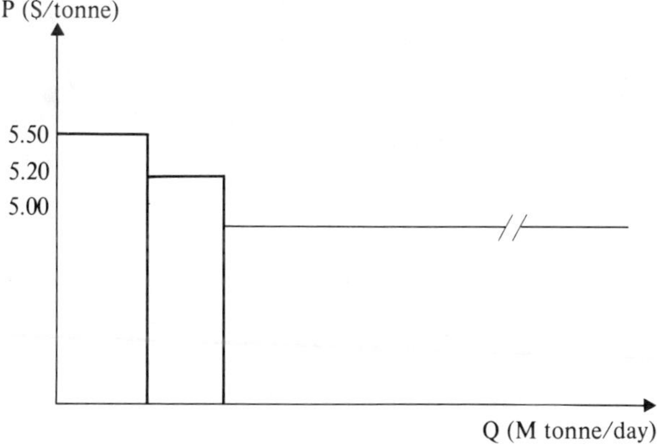

Figure 7-6 Price–quantity curve for ore-processing example.

cost the single investment of a large amount of money at the beginning of operation must be represented on a daily (or yearly) operating basis.

This is accomplished by multiplying the capital cost by a *capital recovery factor* (CRF) (sometimes called an *amortization factor*) which represents the yearly cost of servicing that investment. Given the interest rate i and the number of years n over which the investment capital is to be recovered, the annual capital recovery factor is expressed as:

$$\text{CRF} = \frac{i(1+i)^n}{(1+i)^n - 1} \ (\$/\$/yr) \tag{7-11}$$

This formula can easily be derived as follows.

If P is the capital cost of plant and equipment then R, the annual cost of servicing it, is to be expressed in the form

$$R = (\text{CRF}) \times P \tag{7-12}$$

Now the present value of the series of payments R must return the principal P.

i.e.,

$$\sum_{t=1}^{n} \frac{R}{(1+i)^t} = P \tag{7-13}$$

Let

$$\alpha = \frac{1}{1+i} \tag{7-14}$$

Then Eq. (7-13) becomes

$$R\alpha(1 + \alpha + \alpha^2 + \cdots + \alpha^{n-1}) = P \tag{7-15}$$

Using the summation formula for a geometric progression, Eq. (7-15) becomes

$$R\alpha\left(\frac{1-\alpha^n}{1-\alpha}\right) = P \tag{7-16}$$

Thus, the capital recovery factor is given by

$$\text{CRF} = \frac{1-\alpha}{\alpha(1-\alpha^n)} \tag{7-17}$$

which, using Eq. (7-14), becomes

$$\text{CRF} = \frac{i(1+i)^n}{(1+i)^n - 1} \tag{7-18}$$

Tabulations of CRF for various ranges of i and n are widely available. To reduce it to a daily cost, the CRF is usually simply divided by 365 (although the answer is not exactly the same as using a daily interest rate for i and correspondingly taking n as the total number of days). If the plant does not operate the full year; for instance, it may have regularly scheduled maintenance work, then we divide by the total number of *stream* days (i.e., operating days), to reduce it to a *per stream day* basis.

If P_0 is the capital cost of plant of capacity Q_0 and x_i is a variable representing capacity, Q, then

$$c_i x_i = \text{CRF } P_0\left(\frac{x_i}{Q_0}\right) \tag{7-19}$$

so that

$$c_i = \text{CRF}\left(\frac{P_0}{Q_0}\right) \tag{7-20}$$

This linear relationship is usually valid for only a limited range of x_i about Q_0, since economies of scale tend to make costs vary according to the (0.6–0.7)th power of capacity.

The use of Eq. (7-19) is equivalent to using the straight line joining the point (P_0, Q_0) to the origin as the linear cost function. A better linear approximation is made using a first-order Taylor's series expansion about the point (P_0, Q_0):

$$P \approx P_0 + \frac{\mathrm{d}P}{\mathrm{d}Q}(Q_0) \times (Q - Q_0) \tag{7-21}$$

Now if the cost–capacity relationship is of the form

$$P = aQ^b \qquad b = 0.6\text{–}0.7 \tag{7-22}$$

we have

$$\frac{\mathrm{d}P}{\mathrm{d}Q} = abQ^{b-1} \tag{7-23}$$

Substituting Eq. (7-23) in Eq. (7-21), we have

$$\begin{aligned} P &\approx P_0 + abQ_0^{b-1}(Q - Q_0) \\ &= (P_0 - abQ_0^b) + (abQ_0^{b-1})Q \end{aligned} \tag{7-24}$$

Thus, if x_i is used to represent Q, we have

$$c_i = (abQ_0^{b-1})\,(\text{CRF}) \tag{7-25}$$

and we keep (CRF) $(P_0 - abQ_0^b) = (1 - b)P_0$ (CRF) as a constant to be added to the objective function row once a solution is obtained.

3. Cost of resources The application is straightforward and the same as for component 1 (value of the product) above; if x_i is the amount of resource used and c_i is the cost per unit, then we get a term $-c_i x_i$ in the objective function. (Note that we are maximizing the objective function.) Note that the term c_i is the *variable* cost; *fixed* costs would give constant terms in the objective function, and thus are excluded from the model (although they may be added in once a solution to the model is obtained).

Again, we can express different availabilities at different prices, as shown in Fig. 7-7. The quantities x_1, x_2, and x_3 will have upper bounds placed on them. The solution will take the cheapest first, so x_1 will go to its upper bound before x_2 increases from zero; there is no danger of getting a nonsensical answer.

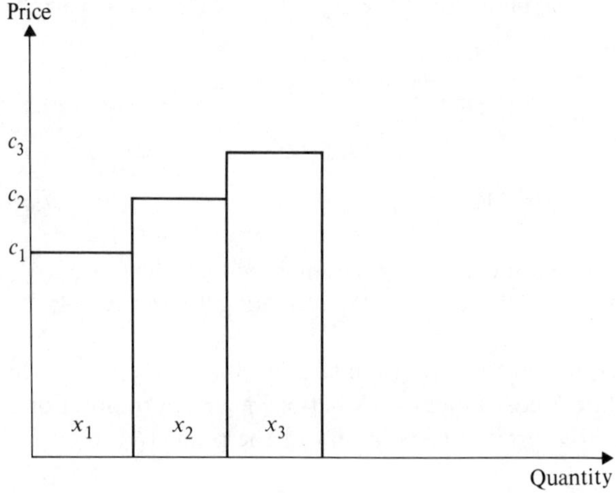

Figure 7-7 Price–quantity curve for resource availability.

4. Operating and maintenance costs These costs are usually a function of the size of plant and equipment, so they could be included with the amortized capital cost. We must consider such items as:

Labor.
Utilities – steam, electricity, water, compressed air, process heat.
Royalties.
Catalyst and other operating requirements.

Example As discussed in Sec. 7-2, we wish to investigate various options for expansion of present units and construction of new units to maximize net profit. The approach we will take is as follows:

1. Enter the size of each unit under investigation into the LP matrix as a capacity constraint.
2. Maximize the operating profit for the stated capacities.
3. Calculate the investment costs for the stated capacities separately from the LP calculations and subtract them from the operating profit.
4. Make parametric adjustments to the capacities and repeat from step 1.

The objective row will be expressed in units of thousands of dollars per stream-day, so if x_i has units of MBSD, then c_i will have units \$/bbl.

We shall retain the convention of maximizing the objective row, so that we enter positive coefficients for sales and negative coefficients for costs. All commercial LP codes allow one to specify whether to maximize or minimize the objective.

1. ***Crude unit***

Bolivian crude costs	12.80 $/bbl
Columbian crude costs	11.48 $/bbl

Thus the objective row will appear as follows:

	VCRDBOL	VCRDCOL
PROFIT	−12.80	−11.48

2. ***Unifiner*** A cost for catalyst of 0.0176 $/bbl of feedstream appears in the objective row.

3. ***Reformer reactor*** Catalyst costs 0.1312 $/bbl feed at 90 RON severity and 0.28 $/bbl feed at 95 RON severity.

4. ***Hydrocracker*** Catalyst costs 0.620 $/bbl feed.

5. ***FCCU*** A catalyst cost of 0.2112 $/bbl feed appears in all columns representing feedstreams to the FCCU.

6. ***Alkylation plant*** Costs for acid and caustic are 0.512 $/bbl of C_3 alkylate and 0.472 $/bbl of C_4 alkylate.

7. ***Utilities*** The costs of utilities are as follows:

	$	Column name
Power, per kw-hr	0.04	PURCKWH
Cooling water, per Mgal.	0.0528	PURCH20
Purchased fuel gas, per MM BTU	1.60	PURCFUL
Steam, per MSCF	0.16	UTILSTM
Purchased IC_4, per bbl	12.00	PURCPC4

Note the column names match those mentioned when discussing the utility balance rows, and their units will correspond to those above; thus, for example, the UBALKWH row will have units Mkw-hr per stream-day, and so will the variable x_{PURCKWH}. c_{PURCKWH} has units $/kw-hr, so $c_i x_i$ becomes M$ per stream-day, as required.

The purchased IC4 is not really a utility; the column is present because we may need to purchase butane for alkylation. In fact, it is a mixture of 50% isobutane and 50% normal butane, both of which are already represented in the model by the flowstreams IC4 and C4P and their corresponding material balance rows (see Sec. 7-4).

The column PURCPC4 thus has entries of -12.00 in the objective row, -0.5 in the NC4 material balance row, and -0.5 in the IC4 material balance row, since these are the "yields" of NC4 and IC4 per barrel of PURCPC4 purchased.

*8. **Products*** The sales realizations for refinery products are as follows:

	\$/barrel	Column name
Premium gasoline	21.44	SELLPRE
Intermediate gasoline	20.32	SELLINT
Regular gasoline	18.04	SELLREG
JP4	16.80	SELLJP4
Diesel	14.40	SELLDSL
Fuel oil	8.00	SELLRSD
LPG	11.00	SELLLPG

CASE EXERCISES

It is intended that the case exercises given below should be conducted in parallel with the subject material in each chapter of Part II. The first exercise is a contrived example which would be suitable as a fairly simple test problem to practice the techniques discussed.

1. Ore processing

An ore processing plant can process two types of ore: Type A is available up to 50 Mtonne/day at a cost of 2.80 \$/tonne. Type B ore is available up to 75 Mtonne/day at a cost of 2.50 \$/tonne. Both types of ore go through a basic processing unit. There are 3 other units in the plant, and they all have running costs and capacity limitations as follows:

Units	Running costs	Capacity limit
Basic processing	20c/tonne	100 Mtonne/day
Treatment	15c/tonne	25 Mtonne/day
Grinding	10c/tonne	40 Mtonne/day
Refining	15c/tonne	40 Mtonne/day

The *sales figures* are as follows:

Product	Revenue	Demand (max)
1	6.00 \$/tonne	No limit
2	5.00 \$/tonne	60 Mtonne/day
3	4.00 \$/tonne	No limit

The *yields* (tonne/tonne feed) are as follows:

Basic processing			Treatment			
	Type A	Type B		PROC 1	PROC 2	GR1
PROC 1	0.15	0.12	TRET 1	0.15	0.20	0.18
PROC 2	0.10	0.10	TRET 2	0.35	0.38	0.40
PROC 3	0.20	0.15	TRET 3	0.50	0.42	0.42
PROC 4	0.23	0.25				
PROC 5	0.32	0.33				

Grinding			Refining		
	PROC 4	PROC 5		PROC 3	GR2
GR1	0.15	0.10	REF 1	0.45	0.30
GR2	0.20	0.20	REF 2	0.55	0.70
GR3	0.25	0.35			
GR4	0.40	0.35			

Note that each column represents an entering feed stream, so it is possible to construct the flow chart from this information.

Product quality specifications

Products 1 and 3 have none. Product 1 consists of TRET 1 and REF 1
Product 3 consists of TRET 3, REF 2 and GR4

Product 2: % metal oxide $\geqslant$ 55

Blend feed stocks for Product 2:

	TRET 2	TRET 3	REF 1	REF 2	GR3	GR4
% metal oxides	65	60	53	50	45	40

We wish to maximize the net revenue per day.

The following case exercises are somewhat more realistic as they require the reader to go out and collect the data himself.

2. Electricity supply

(i) Collect all data on the current electricity supply system of your state, particularly the operating characteristics of the existing plant and the load–duration curve (see Fig. 7-3) of electricity demand.

(ii) Apply a simple growth rate (say 5% p.a.) to predict the demand in 20 years.

(iii) Formulate an LP model to optimize the mix of generating plant, using the existing plant (infrastructure) to give lower bounds on plant capacities.

3. Liquid-fuels model

Extend the refinery model being discussed to include the possibility of substitute fuels being included in the gasoline blends:

(i) Methanol (15% maximum, by volume)
(ii) Syngas (synthetic gasoline)
(iii) MTBE (Methyl-tert-butyl ether).

4. Energy model

(i) Investigate the sources of energy used in your state, their prices and availabilities.
(ii) Investigate the current demands for energy by end-use function, as illustrated in Fig. 7-2.
(iii) Estimate the future demands in, say, 20 years. Again a simple growth rate could be used, but it would be worth delving a little deeper to see how demand patterns may change.
(iv) Estimate the resource availabilities 20 years hence – this may result in re-running the model for a number of scenarios.
(v) Using the existing infrastructure as lower bounds, formulate a model to optimize the energy supply system for the target year.

CHAPTER

EIGHT

LARGE-SCALE MATRIX CONSTRUCTION

8-1 INTRODUCTION

While it will be evident from the discussion in the previous chapter that the problem is formulated in terms of rows, it turns out that the matrix is constructed in terms of columns. In fact, we shall see in Chapter 9 that the standardized LP data input format requires one to enter the nonzero matrix coefficients column by column, rather than row by row.

A slight hint of why this should be so has already been given in Sec. 7-2, where it was pointed out that in the material balance rows the columns tended to be grouped together for each unit of plant and equipment. Further than this, the material balance row entries can be written down term by term for each column in turn.

This natural grouping of columns and nonzero row entries gives rise to the use of tables: a table of data is constructed for each unit of plant and equipment and each set of specification constraints on the products. As long as one remains consistent in the use of row and column labels, the whole matrix can be constructed by listing all the tables, then all the columns within each table, then all the nonzero row entries within each column.

This chapter will describe how this is done.

8-2 TABULATION OF DATA

Equation (7-4) in our example accompanying Sec. 7-4 illustrates how tables can be constructed. In particular, the feedstreams entering the saturated gas unit and the product streams yielded from that unit are described in detail in that equation.

Note that the two columns represent the feedstreams COLNP and BOLNP; there is a corresponding +1.0 entry in their respective material balance rows and the negative entries in the yieldstream material balance rows represent the yield of product per unit feed. We could construct a table representing the complete saturated gas unit by adding further columns representing the feedstreams 90BBG, 95BBG, HYDBBG (see Fig. 7-5).

The conventions to be followed in constructing tables representing units of plant and equipment are given below:

1. Define a column (j) for each entering feedstream (x_j will be the quantity of feed j). Proceed with steps 2–6 for each column so defined.
2. Place a +1.0 entry in the material balance row for the entering feedstream.
3. For each product yielded from the entering feedstream, enter the coefficient $-a_{ij}$ in its corresponding material balance row, where a_{ij} is the yield of product i per unit of feed j.
4. If a capacity constraint based on feed quantity exists for the unit, place a +1.0 entry in the capacity constraint row. The right-hand side of this row will be the bound on the sum of the input feedstreams.
5. In each resource availability constraint row, enter the coefficient $+a_{ij}$, where a_{ij} is the consumption of resource i per unit of feed j (e.g., the utility requirements in our example).
6. Enter the cost coefficient c_j per unit of feed j in the cost row.

We can prepare similar tables for the products: in fact, we could regard the product blending or fabrication operation as a unit of plant and equipment in which a number of feedstreams enter and only one stream (the product itself) is yielded. In addition to the material balance equations, there will be rows for the specification constraints of the type discussed in Sec. 7-6.

The conventions for construction of the product blending tables are as follows:

1. Define a column (j) for each feedstream entering the product blend (x_j will be the quantity of feed j). Proceed with steps 2–5 for each column so defined.
2. Place a +1.0 entry in the material balance row for the entering feedstream.
3. Place a corresponding −1.0 entry in the material balance row (EVOLPROD, say) for the blended product.
4. For each minimum property requirement of the blended product, enter the coefficient $-p_j$ on the corresponding constraint row (see Eq. 7-10).
5. For each maximum property requirement of the blended product, enter the coefficient $+p_j$ in the corresponding constraint row.
6. Having completed steps 2–5 for each feedstream j, define a column (say b) for the blended product (x_b will be the quantity of blended product). In the column make the following row entries:
   (*a*) Place a +1.0 entry in the material balance row (EVOLPROD) for the blended product.

(*b*) For each minimum property requirement of the blended product, enter the coefficient $+p_b$ in the corresponding constraint row (see Eq. 7-10).

(*c*) For each maximum property requirement of the blended product, enter the coefficient $-p_b$ in the corresponding constraint row.

(*d*) If there are bounds on product demand, place a +1.0 entry in the bounding constraint row (this could alternatively be handled by placing explicit bounds on the variable x_b).

(*e*) Enter the value of the product c_b in the objective row.

Example The tables for our example are presented in Table 8-1. Their layout is quite specific; the reason for this will become apparent in Secs. 8-3 and 8-4. In using tables to construct the matrix, we shall see that we need to be systematic in our row and column labeling. Thus the labels do not coincide exactly with the labels used in the flowchart, Fig. 7-5, but their relationship is indicated in the text, T, portion of the tables. (T = 18 indicates the number of characters of text allowable; also "*" in the first column of a statement indicates the statement is not processed, i.e., is merely a comment.)

The material balance rows have labels prefixed MVOL; the utility balance rows are prefixed UBAL (except the hydrogen balance, which is labeled MSCFHYL); the capacity constraints are prefixed VCAP; and the product quality specifications are prefixed X. The material balance rows for blended products are prefixed EVOL and the objective row is labeled PROFIT.

All the conventions discussed in this section have been adhered to; the table UTIL also includes "yields" of heat from various flowstreams (excess gas streams may be burnt for plant fuel) as well as purchased plant fuel (column FUL).

The capacity constraints discussed in Sec. 7-5, constraint type 3, are shown in the tables, and the product sales constraints (Sec. 7-5, constraint type 4) are rows XPSCPRE and XRSCREG in the three gasoline blend tables.

The blend tables consist largely of specification constraints of the type discussed in Sec. 7-6. Of these, the constraints with row labels XLPR ..., XHPR ..., XTEL ... in the gasoline blend tables will be discussed in detail in Sec. 8-6-8. Note that all of the specification constraints are laid out in the form of Eq. (7-10), and the fuel-oil viscosity constraint discussed in Sec. 7-6 is row XVISRSD in the fuel-oil blend table.

The alkylation plant table (VALK) needs some further explanation. The columns LA3 and LA4 represent the amount of C3 alkylate and C4 alkylate produced respectively, so each have entries of −1.0 in their respective material balance rows. The positive coefficients in rows MVOLIC4, MVOLC3U, MVOLC4U in these columns are the requirements of IC4, C3ENE, and C4ENE per barrel of product. This is the reverse representation from other tables, but is still consistent; it turns out that process-engineering calculations for alkylation are based on product flowstreams rather than feed flowstreams. Columns IC4 and C4U have the effect of putting excess IC4 and C4U into the NC4 flowstream, and similarly the C3U column "yields" an equivalent amount of C3S in the C3S flowstream. Excess gas can also be burnt as plant fuel, and so we have similar

Table 8-1 Data tables for example problem

```
DATA
*
   TABLE  VCRD ,  T=18    $CRUDE YIELDS
*
          *        ' T                      '          BOL            COL
          MVOLBOL   BOLIVIAN CRUDE                     1.0
          MVOLCOL   COLUMBIAN CRUDE                                    1.0
          MVOLLNC   COLNP COLUMBIA NAP                               -.2931
          MVOLLNB   BOLNP BOLIVIAN NAP              -.537
          MVOLSRK   KEROP KEROSINE                  -.131            -.1170
          MVOLSRD   DIESLP SR DIESEL                -.1155           -.0649
          MVOLVBB   BBTMP BOL VAC BTM               -.037
          MVOLVBC   CBTMP COL VAC BTM                                -.1800
          MVOLRCR   AGOP  ATM BTM                   -.0365           -.1233
          MVOLHVO   VACGOP VAC GAS OIL              -.143            -.2217

          UBALKWH   KWH                              .302             .384
          UBALH2O   H2O                              .150             .185
          UBALSTM   STEAM                            .003             .003
          UBALFUL   FUEL                             .0587            .1053
          PROFIT    COST                           -12.80           -11.48
*
*
   TABLE  VSGP ,  T=18    $SATURATE  GAS PLT
*
          *        '    T                  '   LNC            LNB        LHG
          MVOLLNC   COLNPG COL NAPH              +1.0
          MVOLLNB   BOLNPG BOL NAPH                             +1.0
          MVOLB95   95BBG  REFORMER 95
          MVOLB90   90BBG  REFORMER 90
          MVOLLHG   HYDBBG UNICRACKER                                      +1.0
          MVOLC3S   C3P  PROPANE              -.0112          -.0277     -.175
          MVOLNC4   C4P  BUTANE               -.0378          -.0563     -.270
          MVOLLSR   LSRNAP  LSR NAPH          -.1502          -.1990     -.028
          MVOLHSR   HSRNAP  HSR NAPH          -.7953          -.6873
          MVOLIC4   IC4P  ISO BUTANE          -.0099          -.017      -.455
          UBALKWH   KWH                        0.721           .495       .495
          UBALH2O   H2O                        0.185           .209       .209
          UBALSTM   STM                        0.013           .013       .013
          UBALFUL   FUEL                        .0488          .0506      .0448
          VCAPSGP   CAP SAT GAS PLT             1.0            1.0
          PROFIT    COST                        0.0            0.0        0.0
*
*                   SATURATE GAS PLANT - CONTINUED
*
          *         B95        B90
          MVOLB95   1.0
          MVOLB90              1.0
          MVOLC3S -.2836     -.2710
          MVOLNC4 -.3285     -.3289
          MVOLLSR -.0241     -.0255
          MVOLHSR
          MVOLIC4 -.2502     -.2656
          UBALKWH  .495       .495
          UBALH2O  .209       .209
          UBALSTM  .0130      .0130
          UBALFUL  .0506      .0506
          VCAPSGP
          PROFIT    0.0        0.0

*
   TABLE  VH2R,    T=18     $UNIFINER
*
          *         '    T                   '   HSR
          MVOLHSR   HSRNAP HSR NAPH              +1.0
          MVOLRFG   REF FEED                     -1.0
          MSCFHYL   NETH2P   H2 CONSP             .0327
          UBALKWH   KWH                           .793
          UBALH2O   H2O                           .045
          UBALFUL   FUEL                          .094
          PROFIT    COST                        -.0176
```

```
    TABLE VRFF, T=18        $REFORMER FEED
*
*                COMBINE FEED TO REFORMER FROM UNIFINER AND UNICRACKER
*                AND SPLIT RESULT BETWEEN 90 AND 95 SEVERITY UNITS
*
        *          '        T            '       RF1       RF2     HH1      HH2
        MVOLRFG    REF FD EX UNIFINER          1.0       1.0
        MVOLR90    REF FD (90 SEV)            -1.0                -1.0
        MVOLR95    REF FD (95 SEV)                      -1.0              -1.0
        MVOLHHG    REF FD EX UNICRACK                              1.0      1.0
*
    TABLE VRFG, T=18        $ REFORMER  YIELDS
*
        *          '        T          ' R90       R95
        MVOLR90    90 SEV REF FEED        1.0
        MVOLR95    95 SEV REF FEED                  1.0
        MVOLB90    90 SEV C4-            -.0404
        MVOLB95    95 SEV C4-                      -.0588
        MVOLF90    90 SEV C5+            -.8564
        MVOLF95    95 SEV C5+                      -.8145
        MSCFHYL    H2 EX REFORMER        -.8239    -.7689
        UBALKWH                           .792      1.03
        UBALH2O                           .297      .387
        UBALSTM                           .0063     .0081
        UBALFUL                          -.156     -.2112
        VCAPRFG    CAP REFORMER           1.0       1.3
        PROFIT                           -.1512    -.304
*
    TABLE VHOL , T=18    $UNICRACKER
*
        *          '        T          '  LCO       SRD       RCR       HVO
        MVOLLCO    CYCLOP CAT CYCL        1.0
        MVOLSRD    DIESLP SR DIESEL                 1.0
        MVOLRCR    AGOP  ATM BTM                              1.0
        MVOLHVO    VACGOP                                               1.0
        MVOLHHG    NAPH EX UNICRACK    -.6627    -.6627    -.5875    -.5875
        MVOLLHG    GAS  EX UNICRACK    -.2414    -.2414    -.3321    -.3321
        MVOLHCD    C5'S EX UNICRACK    -.2930    -.2930    -.3620    -.3620
        MSCFHYL    H2 CONS UNICRACK     2.3       2.3       2.3       2.3
        UBALFUL    FUEL EX UNICRACK    -.2054    -.2054    -.2054    -.2054
        UBALH2O    H2O  TO UNICRACK     .826      .826      .826      .826
        UBALKWH    PWR  TO UNICRACK   14.61     14.61     14.61     14.61

        VCAPHVO    CAP VACGO                                          1.0
        VCAPHOL    CAP UNICRACKER       1.0       1.0       1.0       1.0
        PROFIT     COST                -.2112    -.2112    -.2112    -.2112
*
    TABLE VCCU,T=18       $FCCU UNIT
*
        *         '        T          ' SRK       SRD       RCR       HVO
        MVOLSRK   KEROC  CAT KERO FD     1.0
        MVOLSRD   DIESLC CAT DIES FD               1.0
        MVOLRCR   AGOC   CAT ATM BTM                         1.0
        MVOLHVO   VACGOC CAT VGO FD                                    1.0
        MVOLNC4   C4P    CAT C4       -.0184    -.0184    -.0185    -.0185
        MVOLC3S   C3ANEP CAT C3       -.0303    -.0303    -.0328    -.0328
        MVOLIC4   IC4P   CAT IC4      -.0564    -.0564    -.0568    -.0568
        MVOLC3U   C3ENEP CAT C3=      -.0655    -.0655    -.0658    -.0658
        MVOLC4U   C4ENEP CAT C4=      -.0780    -.0780    -.0806    -.0806
        MVOLFCG   FCNAP  CAT NAPH     -.4750    -.4750    -.4934    -.4934
        MVOLLCO   CYCOLP CAT CYCLEO   -.3050    -.3050    -.2922    -.2922
        MVOLSLR   SLWRYP CAT SLURY     0.        0.       -.0096    -.0096
        UBALSTM   STEAM               -.0654    -.0654    -.0654    -.0654
        UBALFUL   FUEL                -.2703    -.2703    -.2703    -.2703
        UBALH2O   H2O                 0.632     0.632     0.632     0.632
        UBALKWH   KWH                 0.6807    0.6807    0.6807    0.6807

        VCAPHVO   CAPACITY VAC G.O.                                    1.
        VCAPCCU   CAPACITY FCCU       1.        1.        1.          1.
        PROFIT                        -.2112    -.2112    -.2112     -.2112
```

Table 8-1 continued.

```
*
    TABLE VALK, T=18      $ALKYLATION PLANT
*
         *        '    T            ' LA3     LA4      IC4 C3U C4U
         MVOLIC4   ISO-BUTANE'        +.7600  +.6571   +1
         MVOLC3U   PROPYLENE          +.5714               +1
         MVOLC4U   BUTYLENE                   +.5714           +1
         MVOLC3S   PROPANE                    -.0571       -1
         MVOLNC4   N-BUTANE                   -.0114   -1      -1
         MVOLLA3   C3 ALKYLATE        -1.0
         MVOLLA4   C4 ALKYLATE                -1.0
         UBALSTM                       .1869   .1724
         UBALFUL                       .2796   .2579
         UBALH2O                      2.241   2.067
         UBALKWH                      2.766   2.552
         VCAPALK                       1.0     1.0
         PROFIT                        -.512   -.472
LOG,PAGE
*
    TABLE UTIL, T=18        $FUEL AND STEAM BALANCE
*
         *        '   T      ' C3S    NC4     IC4     C3U     C4U     HYL    STM
         MVOLC3S                1.0
         MVOLNC4                        1.0
         MVOLIC4                                1.0
         MVOLC3U                                        1.0
         MVOLC4U                                                1.0
         MSCFHYL                                                        1.0
         UBALSTM                                                               -1.0
         UBALFUL              -3.814 -4.316  -4.153  -3.808  -4.44  -.305   1.42
         PROFIT                                                                -.16
*
    TABLE PURC,   T=18      $MISCELLANEOUS PURCHASES
*                       PC4 - PURCHASE 50-50 MIX OF IC4-NC4
*
         *         '  T    ' PC4          H2O         KWH        FUL      FLR
         MVOLIC4             -.5
         MVOLNC4             -.5
         UBALFUL                                                 -1.0     1.0
         UBALH2O                          -1.0
         UBALKWH                                      -1.0
         PROFIT           -12.0           -.0528     -.04        -1.6
*
    TABLE  LIMT, T=18          $ LIMITS ON MATERIAL BALANCE
*
         *         '     T     ' MAX
   .     MVOLBOL                26.316
         MVOLCOL                21.052
         VCAPSGP                23.26
         VCAPHVO                 5.25
         VCAPRFG                13.455
         VCAPHOL                 3.87
         VCAPCCU                 7.26
         VCAPALK                10.0
*
*                   UNIT TABLES COMPLETE
```

```
*                BEGIN BLEND TABLES
*
    TABLE BPRE , T=18          $ PREMIUM BLEND TABLE
*
         *        LSR     HCD     F95     F90     FCG
         MVOLLSR  1.0
         MVOLF95                  1.0
         MVOLF90                          1.0
         MVOLHCD          1.0
         MVOLFCG                                  1.0
         XLPRPRE  -7.95   -8.84   -9.43   -9.03   -9.23
         XHPRPRE  -8.70   -9.45   -9.57   -9.32   -9.22
         XTELPRE  -3.00   -3.00   -3.     -3.0    -3.
         XRVPPRE  14.00   12.00    3.5     3.5     6.
         X200PRE   1.      1.      .233    .205    .381
         X230PRE  -1.     -1.     -.358   -.333   -.509
         EVOLPRE  -1.     -1.     -1.     -1.     -1.
*
*                PREMIUM BLEND - CONTINUED
*
         *        LA3     LA4     NC4     TEL                 PRE
         MVOLLA3  1.0
         MVOLLA4          1.0
         MVOLNC4                  1.0
         XLPRPRE  -9.4    -9.74   -9.74   -.493               10.03
         XHPRPRE  -9.85   -10.1   -9.9    -.165               10.03
         XTELPRE  -3.     -3.     -3.      1.
         XRVPPRE  2.5      3.3    66.                         -9.5
         X200PRE  .39     .233    1.                          -.5
         X230PRE  -.77    -.58    -1.                         .5
         XPSCPRE                                              .64
         XRSCREG                                              .35
         EVOLPRE  -1.     -1.     -1.                         1.0
         PROFIT                           -.3696             21.44
*
    TABLE BINT, T=18           $ INTERMEDIATE BLEND TABLE
*
         *        LSR     HCD     F95     F90     FCG     NC4     TEL     INT
         MVOLLSR  1.
         MVOLHCD          1.
         MVOLF95                  1.
         MVOLF90                          1.
         MVOLFCG                                  1.
         MVOLNC4                                          1.
         XLPRINT  -7.98   -8.87   -9.46   -9.06   -9.26   -9.77   -.435   9.65
         XHPRINT  -8.58   -9.33   -9.45   -9.20   -9.13   -9.78   -.208   9.65
         XTELINT  -3.0    -3.     -3.     -3.     -3.     -3.     1.
         XRVPINT  14.0    12.     3.5     3.5      6.     66.             -9.5
         X200INT  1.      1.      .233    .205    .318    1.              -0.5
         X230INT  -1.     -1.     -.358   -.333   -.509   -1.             0.5
         XPSCPRE                                                          -.36
         XRSCREG                                                          0.35
         EVOLINT  -1.     -1.     -1.     -1.     -1.     -1.             1.
         PROFIT                                                  -.3696 20.32
```

Table 8-1 continued.

```
      TABLE BREG, T=18      $ REGULAR BLEND TABLE
*
            *        LSR         HCD         F95         F90         FCG
            MVOLLSR   1.
            MVOLHCD               1.
            MVOLF95                           1.
            MVOLF90                                       1.
            MVOLFCG                                                   1.
            XLPRREG  -7.99       -8.88       -9.47       -9.07       -9.27
            XHPRREG  -8.59       -9.34       -9.46       -9.21       -9.14
            XTELREG  -3.         -3.         -3.         -3.         -3.
            XRVPREG  14.         12.          3.5         3.5         6.
            X200REG   1.          1.           .233        .205        .318
            X230REG  -1.         -1.          -.358       -.333       -.509
            EVOLREG  -1.         -1.         -1.         -1.         -1.
            *        NC4         TEL         REG
            MVOLNC4   1.
            XLPRREG  -9.78       -.426        9.05
            XHPRREG  -9.79       -.204        9.05
            XTELREG  -3.         1.
            XRVPREG  66.                     -9.5
            X200REG   1.                     -.5
            X230REG  -1.                      0.5
            XPSCPRE                          -.36
            XRSCREG                          -.65
            EVOLREG  -1.                      1.0
            PROFIT               -.3696      18.04

*
      TABLE BLPG  , T=18        $BLEND LPG
*
            *         '   T    '        C3S         NC4      LPG
            MVOLC3S                     1.0
            MVOLNC4                                 1.0
            EVOLLPG                    -1.0        -1.0      +1.0
            PROFIT                                           11.0
*
      TABLE BJP4  , T=18        $BLEND JET FUEL
*
            *             '  T   '      LSR         HSR          JP4
            MVOLLSR                     1.0
            MVOLHSR                                 1.0
            EVOLJP4                    -1.0        -1.0          1.0
            XRVXJP4                    +14.        +.8          -3.0
            XRVNJP4                    -14.        -.8          +2.0
            PROFIT                                              16.8
*
      TABLE BDSL   , T=18       $BLEND DIESEL
*
            *            '  T  '  SRK         SRD        DSL
            MVOLSRK                1.0
            MVOLSRD                           1.0
            EVOLDSL               -1.0       -1.0        1.0

            PROFIT                                      14.40
*
      TABLE BRSD   ,T=18        $BLEND BUNKER FUEL
*
            *       VBB    VBC    RCR    HVO    SLR    LCO    RSD
            MVOLVBB 1.0
            MVOLVBC        1.0
            MVOLRCR               1.0
            MVOLHVO                      1.0
            MVOLSLR                             1.0
            MVOLLCO                                    1.0
            EVOLRSD -1.0   -1.0   -1.0   -1.0   -1.0   -1.0   +1.0
            XVISRSD 10.1   12.63  6.9    8.05   8.05   4.4    -10.1

            VCAPHVO                      1.0
            PROFIT                                            8.00
*
*                  BLEND TABLES COMPLETE
```

columns appearing in the table UTIL, as mentioned earlier. As we shall see, x_{VALKIC4} will be the amount of excess IC4 blended into the NC4 stream, while x_{UTILIC4} will be the amount of excess IC4 burnt as plant fuel; they are two distinct variables.

8-3 LIST AND TABLE PROCESSING

The nonzero entries in the tables presented in the previous section become the nonzero entries of the LP matrix, and what remains now is to generate the matrix column by column from these tables.

The reader will observe the following three points from the example in the previous section:

1. The rownames for the same constraint row must match even though they are in separate tables. In particular, the material balance rows for each flowstream will always have entries in more than one table.
2. The nomenclature used for each rowname determines what type of row it is (LE$\leqslant$, EQ$=$, and GE$\geqslant$) and whether or not the right-hand side is zero. We shall come back to this point later.
3. Each column in each table has a three-letter label, and each table has a four-letter label. We can therefore construct a seven-letter matrix column name by combining the two labels: each column name is unique and moreover indicates which table it came from.

The procedure to be adopted is as follows:

(*a*) Compile a list of table names. We could possibly have separate lists for each type of table. (In fact in our example we will do this.)
(*b*) Systematically work through the list; for each table on the list generate a matrix column name corresponding to each column in the table.
(*c*) In each column in turn, the nonzero entries, together with their rowname identifiers, become the nonzero matrix entries.

The right-hand side is processed in the same way, as if it were another column of the matrix. Most of the right-hand side entries are zero, as will be recalled from our discussion in Secs. 7-4–7-6. The nonzero entries, for example those associated with capacity constraints, are usually specified in a separate table of their own. (In our example they are specified in the table LIMT.)

Unfortunately, that is not all there is to it. We are also required to specify a list of all the rownames in the matrix. We must also specify what type of row (LE, EQ, GE) each one is. The standard input format for data entry (discussed in Sec. 9-2) requires that in fact all rownames and types must be specified before the nonzero entries of the matrix are inserted. The procedure is as follows:

(*a*) Work through the list of tables.

(*b*) For each rowname in each table, check if it is already on the list of rownames. (Since the same rowname often appears in more than one table, this is quite likely.) If not, add the rowname to the list.

(*c*) Specify the row type, following the convention devised when generating the rownames.

Example The lists of table names are given in Table 8-2.

Table 8-2 Lists of table names

```
*                        BEGIN LIST DEFINITION
*
    LIST (UNIT),T=23
*
          VCRD '      CRUDE UNIT          '
          VSGP '  SATURATE GAS PLANT      '
          VH2R '       UNIFINER           '
          VRFF '   PLATFORMER FEED        '
          VRFG '      PLATFORMER          '
          VHOL '      UNICRACKER          '
          VCCU '   FLUID CAT CRACKER      '
          VALK '   ALKYLATION PLANT       '
          PURC '                          '
          UTIL '                          '
*
    LIST (BLND),     T=24
*
          BPRE          'GASOLINES: PREMIUM          '
          BINT          '           MEDIUM           '
          BREG          '           REGULAR          '
          BLPG          'LPG                         '
          BJP4          'JP4                         '
          BDSL          'DIESEL                      '
          BRSD          'RESIDUUM                    '
*
    LIST (SLS)
*
*                        BUILD SALES LIST - LAST 3 CHARACTERS IN LIST (BLND)
*                             PRE TO RSD
*
*
          (XXX) ,FOR (XXX) = (BLND)* 'ABCD'/ 'BCD'
```

Note that the lists themselves have labels, so that we can refer to each particular class of tables individually. For example, the list (UNIT) is a list of all the units of plant and equipment.

The row types follow the conventions we have adopted for this problem, as shown in Table 8-3.

Table 8-3 Convention in row-labeling for example problem

Row prefix	Row type	Description
MVOL---	=EQ	Material balance rows (except MVOLBOL) MVOLCOL)
X ------	≤LE	Maximum constraints
E ------	=EQ	Equality constraints
VCAP --	≤LE	Capacity constraints
MSCF ---	=EQ	H_2 balance row
UBAL ---	=EQ	Utility balance rows

The right-hand side of all these rows is zero unless the rownames are listed in table LIMT of Table 8-1, in which case their right-hand sides are as given in that table.

8-4 MATRIX GENERATOR LANGUAGES

It will be seen that the work involved in Sec. 8-3 is tedious to say the least. This work is made easy and systematic with the use of matrix generator languages. Such languages are high level, like FORTRAN, and are especially devised to generate the data required for matrix definition from the lists and tables by the use of a series of simple statements.

For the discussion to have any meaning, we will again have to be specific; and we will discuss how the matrix is generated using the language GAMMA. Other languages, such as PDS/MAGEN, have similar capabilities and they are all constantly being upgraded to improve their ease of use and range of options.

An important feature is the ability to use a list label in a statement as if it were an ordinary variable, the implication being that every element on the list gets processed by the statement. Thus, for example, we can process every table in the list (UNIT) of Table 8-2 in one statement without having to refer to each of the table names individually. (Brackets distinguish a list from an individual element.)

Another feature is the ability to refer to elements of tables by the expression (T, R, C) where T is the table name, R is the rowname, and C is the column label. Each of the terms T, R, and C can themselves be list labels, so the statement gets repeated for every member of the list.

Generation of the matrix from the data occurs in two distinct phases: row definition and column definition.

Row definition Commercial codes require the rowname, type, and right-hand side value.

(*a*) The row declaration statement is of the form ⟨^{5}row lab⟩, ⟨type⟩
where ⟨row lab⟩ is the rowname (or better still the label of the *list* of rownames)
and ⟨type⟩ is one of the following symbols:
MAX (or L), i.e., less than or equal to
FREE (or N), i.e., free constraint
MIN (or G), i.e., greater than or equal to
FIX (or E), i.e., equality constraint
OBJ (or N), i.e., objective row

The superscript (5) identifies the column number (on 80 column card format) where the word starts.

(*b*) The right-hand side statement is of the form
⟨^{10}rhslab⟩, ⟨type⟩ = ⟨expression⟩
where ⟨rhslab⟩ is the column-name given to the right-hand side, ⟨type⟩ is

usually RHS, although some other options will be discussed in Sec. 8-6, and ⟨expression⟩ is the numerical value of the right-hand side element. If ⟨expression⟩ is 0.0, then the entire statement can be omitted.

Examples of these statements are given in Table 8-5.

Column definition Having defined the rows, the statement MATRIX precedes the statements defining the column entries.

(*a*) Column declaration statement is of the form
⟨^{5}collab⟩ – which is the column name (j)
(*b*) Coefficient statement is of the form
⟨^{10}rowlab⟩ = ⟨expression⟩
where ⟨rowlab⟩ is the rowname (i) and ⟨expression⟩ is the numerical value of the coefficient (a_{ij}). This statement is repeated for all the nonzero row entries in each column.
(*c*) Bounds statement is of the form
⟨^{10}boundlab⟩, ⟨type⟩ = ⟨expression⟩

This statement puts a bound on the variable ⟨collab⟩ (x_j). ⟨boundlab⟩ is the name given to the bound (not the same as ⟨collab⟩), and ⟨type⟩ is one of:

FREE: x_j can vary between $\pm\infty$, otherwise the implied lower bound is 0.0 (discussed in Sec. 8-6-4)
FIX: x_j is fixed at the value of ⟨expression⟩ (Sec. 8-6-6)
MAX: x_j must be less than or equal to ⟨expression⟩
MIN: x must be greater than or equal to ⟨expression)

It is not necessary to have a bounds statement if the variable x_j must be simply greater than or equal to zero.

It would appear that using these statements to generate the matrix is an even more tedious task than that of Sec. 8-3. However, the use of list labels rather than individual rownames and column names means that it can be accomplished in a very few statements.

Without attempting to give the reader a course on the GAMMA language, it will still be instructive to go through the required statements for our example to illustrate how few statements are actually required.

Example The examples given in Secs. 8-2 and 8-3 have been laid out in the format required by the GAMMA language. The data definition phase is complete when all the lists are defined. This is accomplished by the statements shown in Table 8-4.

Table 8-4 List definition for example problem

```
*                  CREATE LIST OF ALL ROWNAMES BY TYPE
*                       MVOL - FLOW ROWS            EG. MVOLNC4 ETC.
*                       X    - MAX CONSTRAINT           XRVPPRE
*                       EVOL - EQUALITY CONSTRAINT      EVOLPRE
*                       VCAP - CAPACITY CONSTRAINTS     VCAPCCU
*                       MSCF - H2                       MSCFH2
*                       UBAL - UTILITIES                UBALKWH
*                  SEARCH UNIT ROWNAMES, ELIMINATING DUPLICATES (CHECK)
*
     LIST (MVOLXXX) ,CHECK
*
          (MVOLDUM) , FOR (MVOLDUM) = ((UNIT),,*)
                    , FOR (MVOL) = ((UNIT),,*)
                    , IF((MVOL/MVOLDUM).IM.MVOL)
*
     LIST (XYYYBLN) ,CHECK
*
          (XDUMBLN) , FOR (XDUMBLN) = ((BLND),,*)
                    , FOR (X) = ((BLND),,*)
                    , IF((X/XDUMBLN).IM.X)
*
     LIST (EYYYBLN) ,CHECK
*
          (EYYYDUM) , FOR (EYYYDUM) = ((BLND),,*)
                    , FOR (E) =         ((BLND),,*)
                    , IF ((E/EYYYDUM).IM. E)
*
     LIST (VCAPXXX) ,CHECK
*
          (VCAPDUM) , FOR (VCAPDUM) = ((UNIT),,*)
                    , FOR (VCAP)    = ((UNIT),,*)
                    , IF((VCAP/VCAPDUM) .IM. VCAP)
*
     LIST (MSCFXXX) ,CHECK
*
          (MSCFDUM) , FOR (MSCFDUM) = ((UNIT),,*)
                    , FOR (MSCF)    = ((UNIT),,*)
                    , IF((MSCF/MSCFDUM).IM. MSCF)
*
     LIST (UBALXXX) ,CHECK
*
          (UBALDUM) , FOR (UBALDUM) = ((UNIT),,*)
                    , FOR (UBAL)    = ((UNIT),,*)
                    , IF(( UBAL/UBALDUM) .IM. UBAL)
*
*                 LISTS COMPLETE
```

For each type of row there is a corresponding list of rownames. Their construction follows the same pattern, so we will discuss only the list MVOLXXX and how it is created:

```
5
LIST (MVOLXXX), CHECK
```

The list label is MVOLXXX, and the parentheses signify it as a list rather than a single variable. The word CHECK prevents duplicate entries of the same rowname.

```
10
(MVOLDUM), FOR (MVOLDUM) = ((UNIT),,*)
         , FOR (MVOL)    = ((UNIT),,*)
         , IF (MVOL/MVOLDUM .IM . MVOL)
```

Table 8-5 Row definition

```
PROBLEM F
    PROFIT, OBJ
*
*
*
    (MVOLXXX),MAX
         BI, RHS= (LIMT  ,(MVOLXXX),MAX)
              ,IF((LIMT  ,(MVOLXXX),MAX)  .NE. EMPTY)
    (XYYYBLN) ,MAX
    (EYYYBLN) ,FIX
    (VCAPXXX) ,MAX
         BI, RHS= (LIMT  ,(VCAPXXX),MAX)
              ,IF((LIMT  ,(VCAPXXX),MAX)  .NE. EMPTY)
    (MSCFXXX) ,FIX
    (UBALXXX) ,FIX
```

Rather than name the elements individually, as was done for the table names (see the example in Sec. 8-3), the elements of the list MVOLXXX are defined as the elements of the list MVOLDUM, which is created as follows: A temporary list of the same name (MVOLDUM) is made consisting of all seven characters in every rowname in every table in the list (UNIT) (Table 8-2) – this is accomplished by the statement: , FOR (MVOLDUM) = ((UNIT), , *). (The right-hand side of this statement indicates we should go through all rownames in the table (UNIT); but this in turn is a *list* of tables.)

Similarly, a temporary list (MVOL) is made consisting of the first four characters in every rowname in every table in the list UNIT :, FOR (MVOL) = ((UNIT), , *).

The element of the temporary list (MVOLDUM) is added to the (permanent) list MVOLDUM if its corresponding element on the temporary list (MVOL) consists of the four characters MVOL. This is accomplished by the statement , IF ((MVOL/MVOLDUM) . IM . MVOL).

The result of all this is that only the rownames starting with the characters MVOL are added to the list MVOLXXX.

The matrix generation phase comes next. The row definition consists of the statements given in Table 8-5.

By using the lists of rownames rather than each one individually, all the rows are declared with seven row declaration statements and two right-hand side statements. The column name given to the right-hand side is BI, and numerical values of the right-hand sides are in table LIMT (Table 8-1), row (MVOLXXX), i.e. the list MVOLXXX, and column MAX. Note that empty entries in column MAX are suppressed.

The columns definition can be accomplished in only thirteen statements, as shown in Table 8-6.

Again the great power of using lists is demonstrated. A temporary list (COL) is created consisting of all the column labels in all the tables in the list UNIT. Similarly a temporary list (ROWNAME) is created out of all the rownames. The list UNIT is combined with the list COL in the column-declaration statement which defines the column names. In each column so created, all the nonzero row entries are entered in the coefficient statement, using the list ROWNAME and the

Table 8-6 Column definition

```
*
    (UNIT)(COL)        ,FOR (COL)= ((UNIT),*,)
                       ,FOR (ROWNAME) = ((UNIT),,*)
          (ROWNAME) = ((UNIT),(ROWNAME),(COL))
                ,   IF (((UNIT),(ROWNAME),(COL)) .NE. EMPTY)
DO (BLND)              ,FOR (COL)= ((BLND),*,)
                       ,FOR (ROWNAME) = ((BLND),,*)
    (BLND)(COL)        ,IF((SLS/BLND).NE. (COL))         $SUPPRESS SALES COL
          (ROWNAME) = ((BLND),(ROWNAME),(COL))
                ,   IF (((BLND),(ROWNAME),(COL)) .NE. EMPTY)
    SELL(COL)          ,IF((SLS/BLND) .EQ. (COL))
          (ROWNAME) = ((BLND),(ROWNAME),(COL))
                ,   IF (((BLND),(ROWNAME),(COL)) .NE. EMPTY)
END DO.
```

numerical entries in (T, R, C,), again using lists for table, row, and column position.

The same technique is used for the blend tables, but since we wish to distinguish variables representing product sales by column names starting with SELL, we use a DO-loop to repeat the pattern for every table in the list BLND.

Thus the whole matrix is generated using around 22 statements once the lists have been defined.

8-5 ERROR CHECKING

It is a sad fact of life that it usually takes more than one pass at the model to get a working solution, mainly through human failure to get the right numbers and put them in the right place. If a matrix generator language is used, there are less likely to be any errors, but if errors do occur they are more systematic and therefore harder to find.

There are a few error checks that can be made a priori, as follows:

1. Ensure that there is at least one entry apart from the objective row entry, in each column of the matrix. This is easy to check since the data is read in column-wise.
2. Ensure that there is at least one entry in each row. Also, if the right-hand side (b_i) is zero, there must be at least one coefficient (a_{ij}) greater than zero if GE($\geqslant$) and at least one coefficient (a_{ij}) less than zero if LE($\leqslant$) and at least one of each if EQ($=$) if the variable (x_j) is to have any value other than zero (assuming x_j must be nonnegative).
3. Look at the pattern of nonzeros that occur in the matrix. Many commercial LP codes have a "map" or "picture" option where the matrix is laid out as a rectangular array with a special character printed for each nonzero entry. Usually positive and negative coefficients of different magnitudes have different special characters. This device is very useful, and is like having the LP matrix printout without the actual numbers. However, for large matrices it becomes rather hard to decipher since the whole matrix does not fit on one page of computer printout.

Otherwise, we run the model in the computer and get an answer:

1. The problem is infeasible. This is confounding as there is usually no clue as to which row or rows are causing the infeasibility (usually a chain reaction occurs so the row which is declared infeasible by the LP code is not the culprit). One trick which has been found very useful in practice is to make the equality constraints, particularly the material balance rows, temporarily inequality constraints. For example, if the material balance rows are made ≤ 0.0, rather than $=0.0$, the value of the associated slack variables will indicate where the out-of-balance occurs (and therefore which coefficients need to be checked). We could take this further and allow the rows to be "free" but place a cost on both the positive and negative slack variables (Fieldhouse, 1974). (In our example we may designate the MVOL rows as L($\leq$) with impunity.†)
2. The problem is unbounded. We at least get some information: the entering nonbasic variable which goes unbounded. This provides some clue as to which constraint has been ignored. An unbounded solution may also mean an implied nonnegativity of a variable has been ignored when formulating the problem. This is discussed further in the next section.
3. We get an optimal solution in which $x_j = 0.0, j = 1, \ldots, n$. This can happen all too easily. It usually arises in a situation such as trying to minimize the cost of production (without considering the product sales) and having no lower bound on the production quantity. It is therefore optimal to do nothing at all, for then the cost is zero! The remedy is simple: put in a lower bound on production.
4. We get an optimal solution which is unreasonable. The only thing we can do is check the variables which have unreasonable values and the constraint rows in which they occur. Unreasonable numbers often mean that a row is not expressed in consistent units. If the variables are reasonable but the value of the objective function and/or the reduced costs are unreasonable, we must check the cost coefficients. These simple checks for reasonableness are better done by hand (scanning the output) than by programming special routines into the code. A value of the objective which is too high may suggest some missing constraints, while a low value would suggest examining initially the constraints with high shadow prices.
5. We get an optimal solution which is reasonable. How do we know the solution is still correct? There is no easy answer to that question. In general, we need to verify the model by seeing how it responds to changes in values of the coefficients; ideally comparing it with historical data from the actual system being modeled (but there are many situations where this is not possible). When doubt arises, the only way is the hard way: check out all the rows one by one. (It is better to check out the rows using the known (historical) values for the variables x_j, if possible.)

† In fact Table 8-5 shows these rows as L-type.

8-6 TIPS AND TRICKS

This section is a pot pourri of miscellaneous ideas that have been found useful in practice, and various features available in commercial codes. Most of the ideas are fairly simple, but they are worth knowing about and therefore worth mentioning explicitly. One idea has already been mentioned in the previous section: make equality constraints temporarily inequalities to detect infeasibilities. Other ideas presented below are in much the same spirit of expediency as that.

8-6-1 Severity Change Vectors

As was mentioned in Sec. 7-4, first-order effects of changes in operating conditions (severity) can be modeled. Changes in operating conditions of plant and equipment give rise to changes in the product yieldstreams from the unit. For example, suppose the variable x_{SEV} represented the increased severity of operation (e.g., reaction time) of the hydrocracker. For each unit of x_{SEV} the yield of HYDBBG and UNINAP increases by 0.0045 and 0.0752, the yield of UNIGAP decreases by 0.08, and the cost increases by 0.032. (These coefficients are based on an assumed flowrate for the yieldstreams.) There will be an extra column in the hydrocracker table, with the following nonzero entries:

		SEV
MVOLHHG	Naphtha (UNINAP)	−.0752
MVOLHCD	Gasoline (UNIGAP)	+.08
MVOLLHG	C 5's, etc. (HYDBBG)	−.0045
PROFIT		−.032

8-6-2 Collector Rows

This device in its simplest form is merely a convenient way of expressing sums of certain variables explicitly. For example, we could have a row called ESUMFD with a −1 entry in all the feedstream columns, a +1 entry in a new column called CAP, and right-hand side =0.0. Thus, x_{CAP} is simply the sum of the individual feedstream flowrates. The feed capacity is represented as an explicit variable at the cost of an extra row and column. Note that we have avoided this in our example problem. It is usually used when the capital cost of plant and equipment is to be separated from other costs associated with the feedstreams (such as cost of raw material and running costs). By having a separate variable representing capacity, the (amortized) capital cost can be entered as a coefficient in the corresponding column.

Slightly more complicated is the use of collector rows to calculate certain properties of the combined feed from the properties of the individual feedstreams. The yields of products from the unit are based on assumed values of the properties of the combined feed. We put in a new column representing (actual value − assumed value) for each property, and this is used in the same way as the severity-change vector discussed above, except the value of the corresponding variable is calculated from the collector row.

Let a_i be the increase in yield of product i per barrel of feed per unit change in property (assumed the same for each feedstream j entering the unit). The increased flow in the yieldstream is then given by

$$a_i \sum_j x_j \, \Delta p_{\text{act}}$$

where Δp_{act} is the actual value of the property minus the assumed value (p_{a}) and the summation occurs over all the feedstreams entering the unit. Now ΔP_{act} is given by the relationship

$$\sum_j x_j \, \Delta p_{\text{act}} = \sum_j \Delta p_j x_j$$

where $\Delta p_j = p_j - p_{\text{a}}$ and p_j is the property associated with feedstream j. Thus if we define a new column, with label DEL, the value of x_{DEL} is the quantity $(\sum_j x_j \, \Delta p_{\text{act}})$ and the table with the collector row has the extra entries:

	FEED1		FEEDJ	DEL	RHS
Collector Row	$-\Delta p_1$	$\cdots$	$-\Delta p_j$	1.0	=0.0
MVOLYL1				$-a_1$	
MVOLYL2				$-a_2$	
$\vdots$				$\vdots$	
MVOLYLI				$-a_i$	

Note that material balance rows (MVOLYLI) have the coefficients $-a_i$, thus giving the increased flowrate.

As examples we could cite the property UOPK as affecting yields in the FCCU and N + 2A (naphthenes plus two aromatics) as affecting yields in the reformer reactor, although we will not pursue the matter with our current problem.

8-6-3 Implied Non-negativity of Variables

A sound piece of advice is to have at least one variable associated with each stream in the flowchart, for then the implied nonnegativity of variables ensures that the stream will not flow backwards, i.e., have a negative flow.

Consider the four flowstreams represented in Fig. 8-1.

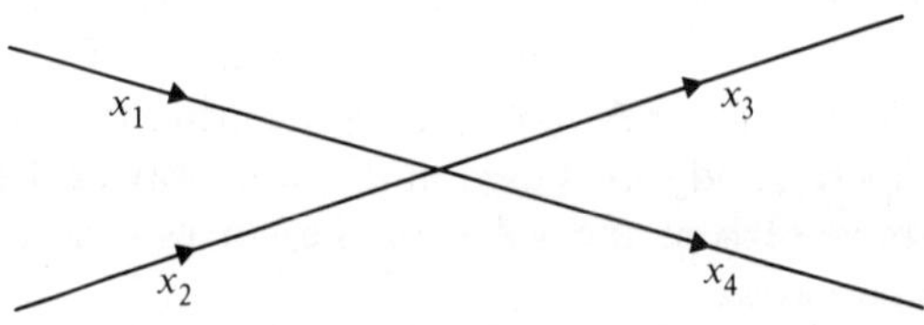

Figure 8-1 Junction of four flowstreams.

This junction has a material balance equation

$$x_1 + x_2 - x_3 - x_4 = 0$$

The stream represented as x_4 is nonnegative since $x_4 \geqslant 0$ by convention. However, if it was represented as $x_1 + x_2 - x_3$ in order to save a row and a column, then if x_3 exceeds $x_1 + x_2$ (and there is nothing stopping this since they are all $\geqslant 0$), then the stream represented by $x_1 + x_2 - x_3$ is negative! This situation is often the cause of an unbounded solution.

In our example we have a variable (column) representing each stream as it enters a process unit or product blending. Note also the junction material balances for the feed entering the reformer reactor (table VRFF in Table 8-1): the two streams RF1 and HH1 enter the junction (which could be regarded as a dummy unit) and the stream R90 leaves. Thus we have

$$-x_{\text{VRFFRF1}} - x_{\text{VRFFHH1}} + x_{\text{VRFGR90}} = 0 \qquad \text{(row label MVOLR90).}$$

The 95RON reactor is analogous.

8-6-4 Free Variables

Most commercial codes allow one to specify "free" variables, which can take any value $-\infty < x_j < +\infty$. They are useful in such areas as material balance rows or resource availability rows where it is not known a priori if the net requirement will be positive or negative.

In our example, the net plant fuel requirements FUL are not known, but rather than use a free variable we have tackled the situation in a slightly different manner.

The UBALFUL utility balance row for plant fuel has positive entries where heat is required and negative entries where heat is yielded. Some of the product streams (IC_4, C_4, C_3, C_3-ene, C_4-ene, H_2) may also be burnt as plant fuel – so each of these products "yields" heat expressed in this row in fuel-oil equivalent terms (see table UTIL in Table 8-1). This row also has a -1.0 entry in the FUL column representing purchased plant fuel, and a $+1.0$ entry in the FLR column representing excess plant fuel which is burnt in the flare (see table PURC in Table 8-1). Thus one of these variables must be zero at the optimal solution; FLR represents the excess and FUL represents the shortage.

8-6-5 Free Rows|Range Rows

We can carry along some extra rows not required for the current run by specifying them as "free" rows. We can also have a *range* on rows, where the right-hand side must be between specified lower and upper bounds. Thus in the rows definition phase of the matrix generator language GAMMA (Sec. 8-4), the right-hand side statement can have the form:

⟨rhslab⟩ , FREE
, MAX = ⟨expression⟩ , MIN = ⟨expression⟩
, MAX = ⟨expression⟩ , RANGE = ⟨expression⟩

We will discuss ranged rows again in Sec. 9-2.

8-6-6 Fixed Variables

Again for a particular run of the model we may wish to fix a variable at a certain value, so that its lower bound is the same as its upper bound. For example, we may wish to evaluate a particular type of crude oil which has become available on the spot market. We put a column representing the crude in the matrix, and make sure it stays nonbasic at zero by "fixing" it. We therefore get the reduced cost of the crude from which we can calculate its price margin.

8-6-7 Pricing Slack Variables

We are not forced to make the cost coefficient of each slack variable zero. We may wish to put in a price representing our estimate of the shadow price of that resource, or perhaps a price which reflects the price margin of increased availability using some other market. We can therefore see directly whether the venture is worth while.

We could also allow a constraint to be *soft* by putting in both a positive and negative slack with appropriate prices.

8-6-8 Nonlinear Response Curves

Often an attribute of a flowstream responds nonlinearly to a certain variable and we are faced with the task of making reasonable linear approximations to the nonlinearity. Separable programming, discussed in Sec. 6-1, can be used but there are certain circumstances under which the special procedures discussed there are not required.

Consider the situation where a property of a product responds to the variable x as shown in Fig. 8-2. The property p is subject to a minimum requirement

$$p \geqslant p_b$$

where p_b is a specified constant.

If we make a first-order Taylor's series approximation to p at the two points (x^0) and (x^1), the minimum property requirement can be expressed as the two constraints

$$p^0 + r(x - x^0) \geqslant p_b$$
$$p^1 + s(x - x^1) \geqslant p_b$$

where r and s are $(\mathrm{d}p/\mathrm{d}x)$ at (x^0) and $(\mathrm{d}p/\mathrm{d}x)$ at (x^1) respectively.

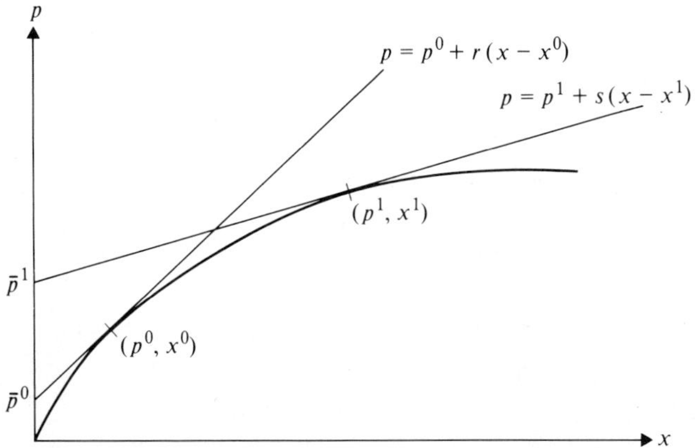

Figure 8-2 Nonlinear response of property p to variable x.

If the response curve is concave (as shown), then the two constraints can be entered individually; since they *both* must be satisfied, we need no special procedure to make each one operative over its appropriate range of x.

The same argument applies for a *maximum* property requirement and a convex response curve. (By the same argument we can also solve convex separable programming problems without special procedures.)

We are not restricted to just two points; first-order approximations can be made at any number of points, giving rise to the same number of constraints. The only danger is that the constraints tend to become linearly dependent as the number of points approximating the same segment of curve increases, so it is usual to have just two or three.

As an example we shall consider the addition of tetra-ethyl lead (TEL) to boost the octane number of the gasoline blends.

It is more convenient to express the constraints in the form:

$$\bar{p}^0 + rx \geqslant p_b \qquad \text{where } \bar{p}^0 = p^0 - rx^0$$
$$\bar{p}^1 + sx \geqslant p_b \qquad \text{where } \bar{p}^1 = p^1 - sx^1$$

$\bar{p}^0$ and $\bar{p}^1$ are the respective intercepts on the p axis in Fig. 8-2.

The blended product must satisfy the relationship

$$\sum_i p_i x_i \geqslant p_b x_b$$

$$\sum_i x_i = x_b$$

where the summation occurs over the product streams (blending stock) entering the blend. Each blending stock has a nonlinear response to tetra-ethyl lead of the form of Fig. 8-2 (where now the variable t will be used in place of x to distinguish it from the blending stocks).

Thus we have the constraints

$$\sum_i (\bar{p}_i^0 + r_i t)x_i \geqslant p_b x_b$$

$$\sum_i (\bar{p}_i^1 + s_i t)x_i \geqslant p_b x_b$$

$$\sum_i x_i = x_b$$

We define the slope of the response curve of the *blend* by the relationship

$$r_b = \sum_i r_i x_i / \sum x_i$$

$$s_b = \sum_i s_i x_i / \sum x_i$$

(This requires an estimate of the proportion of each component in the blend.) If we also define the total TEL used by the variable x_{TEL}, where

$$x_{\text{TEL}} = t \sum_i x_i \text{ (mlTEL/gal)} \qquad \text{(MBSD)}$$

then the constraints become

$$\sum_i \bar{p}_i^0 x_i + r_b x_{\text{TEL}} \geqslant p_b x_b \tag{1}$$

$$\sum_i \bar{p}_i^1 x_i + s_b x_{\text{TEL}} \geqslant p_b x_b \tag{2}$$

$$\sum_i x_i = x_b \tag{3}$$

A further constraint restricts t to 3 ml TEL/gal:

$$x_{\text{TEL}} \leqslant 3.0 \sum x_i \tag{4}$$

Thus, for example, the intermediate gasoline blending table will have the following entries for the four constraints:

	Blending stocks x_i	x_{TEL}	x_b		
XLPRINT	$-\bar{p}_1^0 \ldots -\bar{p}_i^0$	$-r_b$	$+p_b$	$\leqslant 0$	(1)
XHPRINT	$-\bar{p}_1^1 \ldots -\bar{p}_i^1$	$-s_b$	$+p_b$	$\leqslant 0$	(2)
XTELINT	$-3 \ldots -3$	$+1$		$\leqslant 0$	(4)
EVOLINT	$-1 \ldots -1$		$+1$	$=0$	(3)

A glance at the tables (Table 8-1) will verify that this is the way they have been laid out.

CHAPTER

NINE

COMMERCIAL SYSTEMS: ORGANIZATION OF DATA

9-1 INTRODUCTION

This chapter will discuss how the data are prepared for running a model using a commercial linear programming system. The data are entered on cards, or on a tape or disk file of card images. The data can be divided roughly into two classes.

The first is specification of the nonzero matrix elements by row label (i), column label (j), and value a_{ij}. Also the cost row, right-hand side vector, ranges of the right-hand side, and bounds on the variables must be specified. This is discussed in Sec. 9-2. The second class of data is specification of various run-time parameters, such as terminating conditions and error tolerances. These are discussed in Secs. 9-3 and 9-4. Section 9-5 discusses procedures for saving a basis when the current run is terminated, so that subsequent runs can be recommenced from this advanced basis.

9-2 MPS INPUT FORMAT

This format has become the industry standard, adopted by most commercial codes, and is not restricted to the MPS series of codes developed originally for IBM computers.

There is slight variation between codes; some codes allow a wider scope (for example CDC codes allow two more characters in the name fields), so that knowledge of the MPS format is sufficient. The data format is as shown in Table 9-1.

Table 9-1 MPS format data card

	Field 1	Field 2	Field 3	Field 4	Field 5	Field 6
Columns	2–3	5–12	15–22	25–36	40–47	50–61
Contents	Indicator	Name	Name	Value	Name	Value

The various sections of the data are grouped in the following order:

NAME
ROWS
COLUMNS
RHS
RANGES (optional)
BOUNDS (optional)
ENDATA

We will examine each of these sections in turn.

1. *NAME*. This section consists of just one card, with the word NAME in columns 1–4, and the title of the problem in columns 15–22.
2. *ROWS*. In this section all the row labels are defined, as well as the row type. The row type is entered in field 1 (in column 2 or 3) and the row label is entered in field 2 (columns 5–12).

   The code for specifying row type is as follows:

Row type	Indicator (col. 2 or 3)
=	E equality
$\leqslant$	L less than or equal
$\geqslant$	G greater than or equal
Objective	N objective
Free	N no restriction

   This section of data is preceded by a card with ROWS in columns 1–4, followed by a data card for each row. The first N-type row encountered is regarded as the objective, unless it is explicitly identified in the control commands (Sec. 9-3).

   Linear combinations of rows may also be specified. In this case the above row types are denoted respectively by the codes DE, DL, DG, and DN, in columns 2–3. Field 2 contains the linear combination rowname. Fields 3–6 contain the rowname(s) (fields 3 and 5) and their multiplier(s) (fields 4 and 6) which form the combination. A linear combination of three or more rows requires additional cards, following the first card contiguously. In the additional cards field 1 is empty. (The right-hand sides of a linear combination row must be specified in the RHS section, described below.)

3. *COLUMNS*. This section defines the names of the variables, the coefficients of the objective, and all the nonzero matrix elements a_{ij}. The data are entered column by column, and all the data cards for the nonzero entries in each column *must* be grouped together contiguously. The section is preceded by a card with COLUMNS in columns 1–7, followed by data cards which may have one or two matrix elements per card.

   The data card has the column label in field 2 (columns 5–12), the row label in field 3 (columns 15–22), and the value of the coefficient a_{ij} (or c_j) in field 4 (columns 25–36, including a decimal point). If more than one nonzero row entry for the *same* column is to be made on the card, then field 5 (columns 40–47) has the next row label and field 6 (columns 50–61) has its corresponding coefficient value. It should be emphasized that the use of fields 5 and 6 is optional.

   There is no need to specify columns for slack variables; this is taken care of automatically having defined the row types.
4. *RHS*. This section contains the elements of the right-hand side. The section is preceded by a card with RHS in columns 1–3. Since the right-hand side can be regarded as another column of the matrix, the data cards specifying the nonzero entries are in exactly the same format as the COLUMNS data cards, except field 2 (columns 5–12) has a label for the right-hand side. More than one right-hand side may thus be specified in this section; the one to be used for the current run is specified by its label in the agenda cards, to be discussed in Sec. 9.3.
5. *RANGES* (*optional*). This section is for constraints of the following form:

$$h_i \leqslant a_{i1}x_1 + a_{i2}x_2 + \cdots + a_{in}x_n \leqslant u_i$$

   i.e., both an upper and lower bound exist for the row. The range of the constraint is $r_i = u_i - h_i$. The value of u_i or h_i is specified in the RHS section data, and the value of r_i is specified in the RANGES section data. This information, plus the row type specified in the ROWS section, defines the bounds u_i and h_i.

   If b_i is the number entered in the RHS section and r_i is the number specified in the RANGES section, then u_i and h_i are defined as follows:

Row type	Sign of r_i	Lower limit, h_i	Upper limit, u_i
G($\geqslant$)	+ or −	b_i	$b_i + \lvert r_i \rvert$
L($\leqslant$)	+ or −	$b_i - \lvert r_i \rvert$	b_i
E(=)	+	b_i	$b_i + \lvert r_i \rvert$
E(=)	−	$b_i - \lvert r_i \rvert$	b_i

   The section is preceded by a card with RANGES in columns 1–6. The data cards specifying the values of r_i are in exactly the same format as the COLUMNS data cards, except field 2 (columns 5–12) has a label for the column of ranges (which can also be regarded as another column of the matrix). More than one column of ranges may be specified, but all the data cards for each column must be grouped together contiguously.

6. *BOUNDS* (*optional*). In this section, bounds on the variables are specified. The section is preceded by a card with BOUNDS in columns 1–6. The bounds are entered as a row, with a corresponding row label. The nonzero entries in this row vector correspond to columns in the matrix and *must* be in the same order in which the column names appear in the COLUMNS section. When bounds are not specified for a column (or the entire BOUNDS section omitted) the usual bounds, $0 \leqslant x_j < \infty$, are assumed.

   More than one bound for a particular variable may be entered, i.e., both a lower and an upper bound; when only one is specified the other is assumed to be one of the default values of 0 or ∞, as shown in parentheses below.

   Field 1 (columns 2–3) specifies the type of bound:

LO	Lower bound	$b_j \leqslant x_j \; (<\infty)$
UP	Upper bound	$(0 \leqslant) \; x_j \leqslant b_j$
FX	Fixed variable	$x_j = b_j$
FR	Free variable	$-\infty < x_j < +\infty$
MI	Lower bound $-\infty$	$-\infty < x_j \; (\leqslant 0)$
PL	Upper bound $+\infty$ (default bound)	$(0 \leqslant) \; x_j < \infty$

   Field 2 (columns 5–12) specifies the bounds row label. (More than one bounds row may be entered, but the data must be grouped together contiguously for each bounds row.)

   Field 3 (columns 15–22) specifies the column label (j) corresponding to the variable x_j.

   Field 4 (columns 25–36) specifies the bound value b_j.

   Fields 5 and 6 are blank.

7. *ENDATA*. This section is just one card, with ENDATA in columns 1–6, signalling the end of matrix data input.

Example The MPS file for our example problem is shown on the following pages (Table 9-2). (This is the file generated by the GAMMA commands discussed in Sec. 8-4.)

9-3 CONTROL COMMANDS

The MPS file discussed in the previous section may include much extra data which is not required for a particular run. Control commands, often referred to as *agenda cards*, are used to specify various run-time parameters and to specify which portions of the MPS file are to be used for the current run. Most

Table 9-2

```
NAME                OIL REFINERY   EXAMPLE
ROWS
 N  PROFIT
 L  MVOLBOL
 L  MVOLCOL
 E  MVOLLNC
 E  MVOLLNB
 E  MVOLSRK
 E  MVOLSRD
 E  MVOLVBB
 E  MVOLVBC
 E  MVOLRCR
 E  MVOLHVO
 E  UBALKWH
 E  UBALH2O
 E  UBALSTM
 E  UBALFUL
 E  MVOLB95
 E  MVOLB90
 E  MVOLLHG
 E  MVOLC3S
 E  MVOLNC4
 E  MVOLLSR
 E  MVOLHSR
 E  MVOLIC4
 L  VCAPSGP
 E  MVOLRFG
 E  MSCFHYL
 E  MVOLR90
 E  MVOLR95
 E  MVOLF90
 E  MVOLF95
 L  VCAPRFG
 E  MVOLLCO
 E  MVOLHHG
 E  MVOLHCD
 L  VCAPHVO
 L  VCAPHOL
 E  MVOLC3U
 E  MVOLC4U
 E  MVOLFCG
 E  MVOLSLR
 L  VCAPCCU
 E  MVOLLA3
 E  MVOLLA4
 L  VCAPALK
 L  XLPRPRE
 L  XHPRPRE
 L  XTELPRE
 L  XRVPPRE
 L  X200PRE
 L  X230PRE
 E  EVOLPRE
 L  XPSCPRE
 L  XRSCREG
 L  XLPRINT
 L  XHPRINT
 L  XTELINT
 L  XRVPINT
 L  X200INT
 L  X230INT
 E  FVOLINT
 L  XLPRREG
 L  XHPRREG
 L  XTELREG
 L  XRVPREG
 L  X200REG
 L  X230REG
 E  FVOLREG
 E  FVOLLPG
 E  FVOLJP4
 L  XRVXJP4
 L  XRVNJP4
 E  FVOLDSL
 E  FVOLRSD
 L  XVISRSD
COLUMNS
    VCRDBOL       MVOLBOL       1.0
    VCRDBOL       MVOLLNB        -.537
    VCRDBOL       MVOLSRK       -.131
    VCRDBOL       MVOLSRD       -.1155
    VCRDBOL       MVOLVBB       -.037
    VCRDBOL       MVOLRCR       -.0365
    VCRDBOL       MVOLHVO       -.143
    VCRDBOL       UBALKWH       .302
    VCRDBOL       UBALH2O        .150
    VCRDBOL       UBALSTM       .003
    VCRDBOL       UBALFUL        .0587
    VCRDBOL       PROFIT        -12.3
    VCRDCOL       MVOLCOL       1.
    VCRDCOL       MVOLLNC       -.2331
    VCRDCOL       MVOLSRK       -.1170
    VCRDCOL       MVOLSRD       -.0349
    VCRDCOL       MVOLVBC       -.13
    VCRDCOL       MVOLRCR       -.1233
    VCRDCOL       MVOLHVO       -.2217
    VCRDCOL       UBALKWH       .384
    VCRDCOL       UBALH2O       .185
    VCRDCOL       UBALSTM       .003
    VCRDCOL       UBALFUL        .1053
    VCRDCOL       PROFIT        -11.43
    VSGPLNC       MVOLLNC       1.
    VSGPLNC       MVOLC3S       -.0112
    VSGPLNC       MVOLNC4       -.0378
    VSGPLNC       MVOLLSR       -.1502
    VSGPLNC       MVOLHSR       -.7953
    VSGPLNC       MVOLIC4       -.0093
    VSGPLNC       UBALKWH        .721
    VSGPLNC       UBALH2O       .185
    VSGPLNC       UBALSTM        .013
    VSGPLNC       UBALFUL       .0438
    VSGPLNC       VCAPSGP       1.
    VSGPLNB       MVOLLNB       1.
    VSGPLNB       MVOLC3S       -.0277
    VSGPLNB       MVOLNC4       -.0363
    VSGPLNB       MVOLLSR       -.193
    VSGPLNB       MVOLHSR       -.6373
    VSGPLNB       MVOLIC4       -.017
    VSGPLNB       UBALKWH        .495
    VSGPLNB       UBALH2O       .203
    VSGPLNB       UBALSTM        .013
    VSGPLNB       UBALFUL       .0506
    VSGPLNB       VCAPSGP       1.
    VSGPLHG       MVOLLHG       1.0
    VSGPLHG       MVOLC3S       -.175
    VSGPLHG       MVOLNC4       -.27
    VSGPLHG       MVOLLSR       -.028
    VSGPLHG       MVOLIC4       -.435
    VSGPLHG       UBALKWH       .495
    VSGPLHG       UBALH2O        .209
    VSGPLHG       UBALSTM       .013
    VSGPLHG       UBALFUL        .0443
    VSGPB95       MVOLB95       1.
    VSGPB95       MVOLC3S       -.2836
    VSGPB95       MVOLNC4       -.3235
    VSGPB95       MVOLLSR       -.0241
    VSGPB95       MVOLIC4       -.2502
    VSGPB95       UBALKWH       .495
    VSGPB95       UBALH2O        .209
    VSGPB95       UBALSTM       .013
    VSGPB95       UBALFUL        .0506
    VSGPB90       MVOLB90       1.
    VSGPB90       MVOLC3S       -.271
    VSGPB90       MVOLNC4       -.3289
    VSGPB90       MVOLLSR       -.0255
    VSGPB90       MVOLIC4       -.2656
    VSGPB90       UBALKWH       .495
    VSGPB90       UBALH2O        .209
    VSGPB90       UBALSTM       .013
    VSGPB90       UBALFUL        .0506
    VH2RHSR       MVOLHSR       1.
    VH2RHSR       MVOLRFG       -1.
    VH2RHSR       MSCFHYL        .0327
    VH2RHSR       UBALKWH       .793
    VH2RHSR       UBALH2O        .045
    VH2RHSR       UBALFUL       .094
    VH2RHSR       PROFIT        -.0176
    VRFFRF1       MVOLRFG         1.0
    VRFFRF1       MVOLR90       -1.0
    VRFFRF2       MVOLRFG         1.0
    VRFFRF2       MVOLR95       -1.0
    VRFFHH1       MVOLR90        -1.0
    VRFFHH1       MVOLHHG       1.0
    VRFFHH2       MVOLR95        -1.0
    VRFFHH2       MVOLHHG       1.0
    VRFGR90       MVOLR90       1.0
    VRFGR90       MVOLB90       -.0404
    VRFGR90       MVOLF90        -0.8564
    VRFGR90       MSCFHYL        -0.3230
    VRFGR90       UBALKWH       .792
    VRFGR90       UBALH2O       .297
    VRFGR90       UBALSTM       0.0063
    VRFGR90       UBALFUL        -0.136
    VRFGR90       VCAPRFG       1.0
    VRFGR90       PROFIT         -0.1512
    VRFGR95       MVOLR95         1.0
    VRFGR95       MVOLB95       -0.0588
    VRFGR95       MVOLF95        -0.8145
    VRFGR95       MSCFHYL       -.7533
    VRFGR95       UBALKWH        1.03
    VRFGR95       UBALH2O       .387
    VRFGR95       UBALSTM         0.0081
    VRFGR95       UBALFUL       -.2112
    VRFGR95       VCAPRFG         1.3
    VRFGR95       PROFIT         -0.304
    VHOLLCO       MVOLLCO         1.0
    VHOLLCO       MVOLHHG       -.6627
    VHOLLCO       MVOLLHG        -0.2414
    VHOLLCO       MVOLHCD       -.2930
    VHOLLCO       MSCFHYL         2.3
    VHOLLCO       UBALFUL       -.2054
    VHOLLCO       UBALH2O         0.826
    VHOLLCO       UBALKWH       14.61
    VHOLLCO       VCAPHOL         1.0
    VHOLLCO       PROFIT         -0.2112
    VHOLSRD       MVOLSRD         1.0
    VHOLSRD       MVOLHHG       -.6627
    VHOLSRD       MVOLLHG        -0.2414
    VHOLSRD       MVOLHCD       -.2930
    VHOLSRD       MSCFHYL         2.3
    VHOLSRD       UBALFUL       -.2054
    VHOLSRD       UBALH2O         0.826
    VHOLSRD       UBALKWH       14.61
    VHOLSRD       VCAPHOL         1.0
    VHOLSRD       PROFIT         -0.2112
    VHOLRCR       MVOLRCR         1.0
    VHOLRCR       MVOLHHG       -.5875
    VHOLRCR       MVOLLHG        -0.3321
    VHOLRCR       MVOLHCD       -.3620
    VHOLRCR       MSCFHYL         2.3
    VHOLRCR       UBALFUL       -.2054
    VHOLRCR       UBALH2O         0.826
```

Table 9-2 continued.

VHOLRCR	UBALKWH	14.61
VHOLRCR	VCAPHOL	1.0
VHOLRCR	PROFIT	-0.2112
VHOLHVO	MVOLHVO	1.0
VHOLHVO	MVOLHHG	-.5875
VHOLHVO	MVOLLHG	-0.3321
VHOLHVO	MVOLHCD	-.3620
VHOLHVO	MSCFHYL	2.3
VHOLHVO	UBALFUL	-.2054
VHOLHVO	UBALH20	0.826
VHOLHVO	UBALKWH	14.61
VHOLHVO	VCAPHVO	1.0
VHOLHVO	VCAPHOL	1.0
VHOLHVO	PROFIT	-0.2112
VCCUSRK	MVOLSRK	1.0
VCCUSRK	MVOLNC4	-0.0184
VCCUSRK	MVOLC3S	-0.0303
VCCUSRK	MVOLIC4	-0.0554
VCCUSRK	MVOLC3U	-0.0655
VCCUSRK	MVOLC4U	-0.0730
VCCUSRK	MVOLFCG	-0.4750
VCCUSRK	MVOLLCO	-0.3050
VCCUSRK	UBALSTM	-.0554
VCCUSRK	UBALFUL	-.2703
VCCUSRK	UBALH20	.632
VCCUSRK	UBALKWH	.6807
VCCUSRK	VCAPCCU	1.
VCCUSRK	PROFIT	-.2112
VCCUSRD	MVOLSRD	1.
VCCUSRD	MVOLNC4	-.0134
VCCUSRD	MVOLC3S	-.0303
VCCUSRD	MVOLIC4	-.0554
VCCUSRD	MVOLC3U	-.0655
VCCUSRD	MVOLC4U	-.0730
VCCUSRD	MVOLFCG	-.4750
VCCUSRD	MVOLLCO	-.3050
VCCUSRD	UBALSTM	-.0554
VCCUSRD	UBALFUL	-.2703
VCCUSRD	UBALH20	0.632
VCCUSRD	UBALKWH	.6807
VCCUSRD	VCAPCCU	1.
VCCUSRD	PROFIT	-.2112
VCCURCR	MVOLRCR	1.0
VCCURCR	MVOLNC4	-.0145
VCCURCR	MVOLC3S	-.0328
VCCURCR	MVOLIC4	-.0568
VCCURCR	MVOLC3U	-.0653
VCCURCR	MVOLC4U	-.0806
VCCURCR	MVOLFCG	-.4334
VCCURCR	MVOLLCO	-.2322
VCCURCR	MVOLSLR	-.0036
VCCURCR	UBALSTM	-.0554
VCCURCR	UBALFUL	-.2703
VCCURCR	UBALH20	0.632
VCCURCR	UBALKWH	.6807
VCCURCR	VCAPCCU	1.
VCCURCR	PROFIT	-.2112
VCCUHVO	MVOLHVO	1.0
VCCUHVO	MVOLNC4	-.0135
VCCUHVO	MVOLC3S	-.0325
VCCUHVO	MVOLIC4	-.0563
VCCUHVO	MVOLC3U	-.0658
VCCUHVO	MVOLC4U	-.0805
VCCUHVO	MVOLFCG	-.4334
VCCUHVO	MVOLLCO	-.2322
VCCUHVO	MVOLSLR	-.0045
VCCUHVO	UBALSTM	-.0554
VCCUHVO	UBALFUL	-.2703
VCCUHVO	UBALH20	0.632
VCCUHVO	UBALKWH	.6807
VCCUHVO	VCAPHVO	1.
VCCUHVO	VCAPCCU	1.
VCCUHVO	PROFIT	-.2112
VALKLA3	MVOLIC4	.7600
VALKLA3	MVOLC3U	.5714
VALKLA3	MVOLLA3	-1.0
VALKLA3	UBALSTM	.1869
VALKLA3	UBALFUL	.2735
VALKLA3	UBALH20	2.241
VALKLA3	UBALKWH	2.755
VALKLA3	VCAPALK	1.0
VALKLA3	PROFIT	-.512
VALKLA4	MVOLIC4	.6571
VALKLA4	MVOLC4U	.5714
VALKLA4	MVOLC3S	-.0571
VALKLA4	MVOLNC4	-.0114
VALKLA4	MVOLLA4	-1.0
VALKLA4	UBALSTM	.1724
VALKLA4	UBALFUL	.2579
VALKLA4	UBALH20	2.067
VALKLA4	UBALKWH	2.552
VALKLA4	VCAPALK	1.0
VALKLA4	PROFIT	-.472
VALKIC4	MVOLIC4	1.0
VALKIC4	MVOLNC4	-1.0
VALKC3U	MVOLC3U	1.0
VALKC3U	MVOLC3S	-1.0
VALKC4U	MVOLC4U	1.0
VALKC4U	MVOLNC4	-1.0
UTILC3S	MVOLC3S	1.
UTILC3S	UBALFUL	-3.314
UTILNC4	MVOLNC4	1.
UTILNC4	UBALFUL	-4.316
UTILIC4	MVOLIC4	1.
UTILIC4	UBALFUL	-4.153
UTILC3U	MVOLC3U	1.
UTILC3U	UBALFUL	-3.308

UTILC4U	MVOLC4U	1.
UTILC4U	UBALFUL	-4.44
UTILHYL	MSCFHYL	1.
UTILHYL	UBALFUL	-.305
UTILSTM	UBALSTM	-1.
UTILSTM	UBALFUL	1.42
UTILSTM	PROFIT	-.15
PURCPC4	MVOLIC4	-.5
PURCPC4	MVOLNC4	-.5
PURCPC4	PROFIT	-12.
PURCH20	UBALH20	-1.
PURCH20	PROFIT	-.0528
PURCKWH	UBALKWH	-1.
PURCKWH	PROFIT	-.04
PURCFUL	UBALFUL	-1.
PURCFUL	PROFIT	-1.5
PURCFLR	UBALFUL	1.
BLPGC3S	MVOLC3S	1.0
BLPGC3S	EVOLLPG	-1.0
BLPGNC4	MVOLNC4	1.0
BLPGNC4	EVOLLPG	-1.0
SELLLPG	EVOLLPG	1.0
SELLLPG	PROFIT	11.0
BJP4LSR	MVOLLSR	1.0
BJP4LSR	EVOLJP4	-1.0
BJP4LSR	XRVXJP4	14.0
BJP4LSR	XRVNJP4	-14.0
BJP4HSR	MVOLHSR	1.0
BJP4HSR	EVOLJP4	-1.0
BJP4HSR	XRVXJP4	0.8
BJP4HSR	XRVNJP4	-0.8
SELLJP4	EVOLJP4	1.0
SELLJP4	XRVXJP4	-3.0
SELLJP4	XRVNJP4	2.0
SELLJP4	PROFIT	16.8
BDSLSRK	MVOLSRK	1.0
BDSLSRK	EVOLDSL	-1.0
BDSLSRD	MVOLSRD	1.0
BDSLSRD	EVOLDSL	-1.0
SELLDSL	EVOLDSL	1.0
SELLDSL	PROFIT	14.4
BPREFLSR	MVOLLSR	1.
BPREFLSR	XLPRPRE	-7.35
BPREFLSR	XHPRPRE	-8.70
BPREFLSR	XTELPRE	-3.00
BPREFLSR	XRVPPRE	14.00
BPREFLSR	X200PRE	1.
BPREFLSR	X230PRE	-1.
BPREFLSR	EVOLPRE	-1.
BPREFHCD	MVOLHCD	1.0
BPREFHCD	XLPRPRE	-8.34
BPREFHCD	XHPRPRE	-9.45
BPREFHCD	XTELPRE	-3.00
BPREFHCD	XRVPPRE	12.00
BPREFHCD	X200PRE	1.
BPREFHCD	X230PRE	-1.
BPREFHCD	EVOLPRE	-1.
BPREFF95	MVOLF95	1.0
BPREFF95	XLPRPRE	-9.43
BPREFF95	XHPRPRE	-9.57
BPREFF95	XTELPRE	-3.
BPREFF95	XRVPPRE	3.5
BPREFF95	X200PRE	.233
BPREFF95	X230PRE	-.358
BPREFF95	EVOLPRE	-1.
BPREFF90	MVOLF90	1.0
BPREFF90	XLPRPRE	-9.03
BPREFF90	XHPRPRE	-9.32
BPREFF90	XTELPRE	-3.0
BPREFF90	XRVPPRE	3.5
BPREFF90	X200PRE	.205
BPREFF90	X230PRE	-.333
BPREFF90	EVOLPRE	-1.
BPREFFCG	MVOLFCG	1.0
BPREFFCG	XLPRPRE	-9.23
BPREFFCG	XHPRPRE	-9.[illegible]
BPREFFCG	XTELPRE	-3.
BPREFFCG	XRVPPRE	5.
BPREFFCG	X200PRE	.381
BPREFFCG	X230PRE	-.500
BPREFFCG	EVOLPRE	-1.
BPREFLA3	MVOLLA3	1.0
BPREFLA3	XLPRPRE	-9.4
BPREFLA3	XHPRPRE	-9.35
BPREFLA3	XTELPRE	-3.0
BPREFLA3	XRVPPRE	2.5
BPREFLA3	X200PRE	0.33
BPREFLA3	X230PRE	-0.77
BPREFLA3	EVOLPRE	-1.0
BPREFLA4	MVOLLA4	1.0
BPREFLA4	XLPRPRE	-9.74
BPREFLA4	XHPRPRE	-10.1
BPREFLA4	XTELPRE	-3.0
BPREFLA4	XRVPPRE	3.3
BPREFLA4	X200PRE	0.233
BPREFLA4	X230PRE	-0.53
BPREFLA4	EVOLPRE	-1.0
BPRENC4	MVOLNC4	1.0
BPRENC4	XLPRPRE	-9.74
BPRENC4	XHPRPRE	-9.9
BPRENC4	XTELPRE	-3.0
BPRENC4	XRVPPRE	66.0
BPRENC4	X200PRE	1.0
BPRENC4	X230PRE	-1.0
BPRENC4	EVOLPRE	-1.0
BPRETEL	XLPRPRE	-0.433
BPRETEL	XHPRPRE	-0.155

```
BPRETEL    XTELPRE    1.0
BPRETEL    PROFIT     -0.3636
SELLPRE    XLPRPRE    10.03
SELLPRE    XHPRPRE    10.03
SELLPRE    XRVPPRE    -9.5
SELLPRE    X200PRE    -0.5
SELLPRE    X230PRE    0.5
SELLPRE    XPSCPRE    0.64
SELLPRE    XRSCREG    0.35
SELLPRE    EVOLPRE    1.0
SELLPRE    PROFIT     21.44
BINTLSR    MVOLLSR    1.0
BINTLSR    XLPRINT    -7.38
BINTLSR    XHPRINT    -8.58
BINTLSR    XTELINT    -3.0
BINTLSR    XRVPINT    14.0
BINTLSR    X200INT    1.0
BINTLSR    X230INT    -1.0
BINTLSR    EVOLINT    -1.0
BINTHCD    MVOLHCD    1.
BINTHCD    XLPRINT    -8.37
BINTHCD    XHPRINT    -9.33
BINTHCD    XTELINT    -3.0
BINTHCD    XRVPINT    12.0
BINTHCD    X200INT    1.0
BINTHCD    X230INT    -1.
BINTHCD    EVOLINT    -1.0
BINTF95    MVOLF95    1.
BINTF95    XLPRINT    -9.46
BINTF95    XHPRINT    -9.45
BINTF95    XTELINT    -3.0
BINTF95    XRVPINT    3.5
BINTF95    X200INT    .233
BINTF95    X230INT    -.358
BINTF95    EVOLINT    -1.0
BINTF90    MVOLF90    1.
BINTF90    XLPRINT    -9.06
BINTF90    XHPRINT    -9.20
BINTF90    XTELINT    -3.0
BINTF90    XRVPINT    3.5
BINTF90    X200INT    .205
BINTF90    X230INT    -.333
BINTF90    EVOLINT    -1.0
BINTFCG    MVOLFCG    1.
BINTFCG    XLPRINT    -9.26
BINTFCG    XHPRINT    -9.13
BINTFCG    XTELINT    -3.0
BINTFCG    XRVPINT    6.
BINTFCG    X200INT    .313
BINTFCG    X230INT    -.500
BINTFCG    EVOLINT    -1.0
BINTNC4    MVOLNC4    1.
BINTNC4    XLPRINT    -9.77
BINTNC4    XHPRINT    -9.73
BINTNC4    XTELINT    -3.0
BINTNC4    XRVPINT    56.
BINTNC4    X200INT    1.0
BINTNC4    X230INT    -1.
BINTNC4    EVOLINT    -1.0
BINTTEL    XLPRINT    -.435
BINTTEL    XHPRINT    -.203
BINTTEL    XTELINT    1.
BINTTEL    PROFIT     -.3636
SELLINT    XLPRINT    9.65
SELLINT    XHPRINT    9.65
SELLINT    XRVPINT    -9.5
SELLINT    X200INT    -0.5
SELLINT    X230INT    0.5
SELLINT    XPSCPRE    -.36
SELLINT    XRSCREG    0.35
SELLINT    EVOLINT    1.0
SELLINT    PROFIT     21.32
BREGLSR    MVOLLSR    1.0
BREGLSR    XLPRREG    -7.38
BREGLSR    XHPRREG    -8.58
BREGLSR    XTELREG    -3.0
BREGLSR    XRVPREG    14.0
BREGLSR    X200REG    1.0
BREGLSR    X230REG    -1.0
BREGLSR    EVOLREG    -1.0
BREGHCD    MVOLHCD    1.0
BREGHCD    XLPRREG    -8.36
BREGHCD    XHPRREG    -9.34
BREGHCD    XTELREG    -3.0
BREGHCD    XRVPREG    12.0
BREGHCD    X200REG    1.0
BREGHCD    X230REG    -1.0
BREGHCD    EVOLREG    -1.0
BREGF95    MVOLF95    1.0
BREGF95    XLPRREG    -9.47
BREGF95    XHPRREG    -9.46
BREGF95    XTELREG    -3.0
BREGF95    XRVPREG    3.5
BREGF95    X200REG    .233
BREGF95    X230REG    -0.358
BREGF95    EVOLREG    -1.0
BREGF90    MVOLF90    1.0
BREGF90    XLPRREG    -9.07
BREGF90    XHPRREG    -9.21
BREGF90    XTELREG    -3.0
BREGF90    XRVPREG    3.5
BREGF90    X200REG    .205
BREGF90    X230REG    -0.333
BREGF90    EVOLREG    -1.0
BREGFCG    MVOLFCG    1.0
BREGFCG    XLPRREG    -9.27
BREGFCG    XHPRREG    -9.14
BREGFCG    XTELREG    -3.0
BREGFCG    XRVPREG    6.0
BREGFCG    X200REG    0.318
BREGFCG    X230REG    -0.500
BREGFCG    EVOLREG    -1.0
BREGNC4    MVOLNC4    1.0
BREGNC4    XLPRREG    -9.78
BREGNC4    XHPRREG    -9.79
BREGNC4    XTELREG    -3.0
BREGNC4    XRVPREG    56.0
BREGNC4    X200REG    1.0
BREGNC4    X230REG    -1.0
BREGNC4    EVOLREG    -1.0
BREGTEL    XLPRREG    -0.426
BREGTEL    XHPRREG    -.204
BREGTEL    XTELREG    1.0
BREGTEL    PROFIT     -0.3636
SELLREG    XLPRREG    9.05
SELLREG    XHPRREG    9.05
SELLREG    XRVPREG    -9.5
SELLREG    X200REG    -0.5
SELLREG    X230REG    0.5
SELLREG    XPSCPRE    -0.36
SELLREG    XRSCREG    -0.65
SELLREG    EVOLREG    1.0
SELLREG    PROFIT     13.04
BRSDVBB    MVOLVBB    1.0
BRSDVBB    EVOLRSD    -1.0
BRSDVBB    XVISRSD    10.1
BRSDVBC    MVOLVBC    1.0
BRSDVBC    EVOLRSD    -1.0
BRSDVBC    XVISRSD    12.63
BRSDRCR    MVOLRCR    1.0
BRSDRCR    EVOLRSD    -1.0
BRSDRCR    XVISRSD    6.9
BRSDHVO    MVOLHVO    1.0
BRSDHVO    EVOLRSD    -1.0
BRSDHVO    XVISRSD    3.05
BRSDHVO    VCAPHVO    1.0
BRSDSLR    MVOLSLR    1.0
BRSDSLR    EVOLRSD    -1.0
BRSDSLR    XVISRSD    3.05
BRSDLCO    MVOLLCO    1.0
BRSDLCO    EVOLRSD    -1.0
BRSDLCO    XVISRSD    4.4
SELLRSD    EVOLRSD    1.0
SELLRSD    XVISRSD    -10.1
SELLRSD    PROFIT     3.00
RHS
LIMTMAX    MVOLBOL    26.315
LIMTMAX    MVOLCOL    21.052
LIMTMAX    VCAPSGP    23.25
LIMTMAX    VCAPHVO    5.25
LIMTMAX    VCAPREG    13.455
LIMTMAX    VCAPHOL    3.87
LIMTMAX    VCAPCCU    7.25
LIMTMAX    VCAPALK    10.
ENDATA
```

commercial codes also allow the user to write a control program in a high-level language in order to perform such operations as branching and looping to solve a sequence of problems.

Since the command language varies from code to code, we will only be able to talk in general terms about the function of the control commands. The information to be specified is as follows:

1. *Problem specification from the MPS file*
   Maximize or minimize the cost row.
   Cost row label (there may be more than one on the file).
   RHS label (,,).
   Range column label (,,).
   Bounds row label (,,).

2. *Revision options*
   Revise MPS file before processing. Special commands allow one to add, delete, or modify rows or columns or specific matrix elements, without having to reconstruct the entire MPS file.

3. *Input processing options*
   Ignore minor errors.
   Remove redundancies in the specified matrix (Brearley *et al.*, 1975).

4. *Output options*
   "Picture" of matrix (discussed in Sec. 8-5).
   List row and column labels.
   Print out equations (i.e., list coefficients by rows).
   Frequency of iteration log (Sec. 10-1).

5. *Starting options*
   Crash: go through Phase I procedure (Sec. 2-7).
   Advanced basis – from user-supplied file (Sec. 9-5).

6. *Processing options*
   Multiple pricing (Sec. 3-6)
   Partial pricing (Sec. 3-6)
   Scale matrix (Sec. 3-3)

7. *Terminating options*
   Stop after specified number of iterations (regardless of optimality).
   Save basis for subsequent runs (Sec. 9-5).

While the list seems slightly overwhelming, it should be pointed out that most options have default values. For example, the system will process the first RHS in the MPS data if none is specified explicitly, and will terminate after, say, 10,000 iterations if no upper limit is specified.

9-4 ERROR TOLERANCES

Error tolerances were mentioned briefly in Sec. 3-3-4. Here we propose to enumerate the tolerances which can be set. The default value which would be typical for a 60-bit word will be given, as well as some remarks on the effect of changing to a different value.

1. TOLDJ (10^{-5}). This is the reduced cost tolerance. If d_j is greater than $-$TOLDJ for nonbasic variables at zero, or less than $+$TOLDJ for nonbasic variables at their upper bound, then the solution is optimal. If TOLDJ is increased, it may cause premature termination, while if it is decreased it may cause unnecessary extra reinversions and iterations. It is probably better to err on the high side.
2. TOLUR (10^{-3}).
   TOLUC (10^{-1}). These tolerances are used in the P^4 procedure for reinversion (Sec. 3-5). The preassigned order is maintained in order to preserve sparsity unless these tolerances are violated. Any pivot accepted in a column must be no smaller than TOLUR times the largest remaining potential pivot in that column. Otherwise, a replacement column is selected so that its pivot is at least TOLUC times the largest pivot available in that row of the remaining spikes in the bump. Increasing either TOLUR or TOLUC (or both) will increase numerical accuracy at a cost of greater fill-in of the LU factorization.
3. TOLDES (10^{-5}).
   TOLPIV (10^{-8}). These two tolerances are checks on divisors. A divisor smaller than TOLDES is not used if it can be avoided (for example, during suboptimization). If all available pivots are less than TOLPIV, then a reinversion is called for. Increasing the value of these tolerances may give longer total run time for a solution.
4. TOLAIJ (10^{-10}). Any coefficient a_{ij} smaller than TOLAIJ is regarded as zero. This is also often used as an absolute zero check on computed numbers as well.
5. TOLRHS (10^{-6}).
   TOLDIF (10^{-5}). TOLRHS is the tolerance on the row residual ($\mathbf{Ax} - \mathbf{b}$), and TOLDIF is the tolerance on the column residual ($\mathbf{c}_B - \mathbf{B}^{\mathrm{T}}\boldsymbol{\pi}$) (see Sec. 3-3-3). If any residual violates the tolerance, then a reinversion is called for. Increasing the value of these tolerances may give a shorter total run time for a solution.

9-5 GETOFF AND RESTART PROCEDURES

Most commercial codes have one or more procedures for saving a basis on termination of a run. Subsequent runs on the same problem can then be started from the advanced basis, thus saving the necessity of going through all the previous iterations. While each code has its own set of procedures, there is one type of file which has become the industry standard, and this will be described in detail.

When restarting a subsequent run, the MPS-format input data file specifying the matrix (Sec. 9-2) still needs to be entered unless the matrix itself has been saved on another file (e.g., PROBFILE), and the procedures to be discussed specify which variables are in the current basis.

RESTART: Entering an existing basis This is called the INSERT file in the MPS series of codes, and the BASIS file in the APEX series of codes. It allows one to start from an advanced basis, which has been saved from an earlier run, and perhaps modified. The input data are grouped in the following sections:

NAME
Data cards
ENDATA

The NAME and ENDATA cards are analogous to the MPS input file discussed in Sec. 9-2, and the data cards have the same format as MPS input format (Table 9-1).

The function of the data cards is to specify partially or completely the basis. It is assumed that the basis is initially the set of slack (logical) variables, and the structural variables are nonbasic at their lower bound. The data cards then serve to replace systematically the logical variables which are known to be nonbasic, with structural variables which are known to be basic. The data is entered as follows:

Field 1 (*Cols. 2–3*)	*Field 2* (*5–12*)	*Field 3* (*15–22*)
BS	NAME 1	NAME 2

NAME 1 is the column label for a variable
NAME 2 is the column label for a slack variable (which is the row label of its corresponding row).

BS is one of the following four codes specifying basis status:

XL	NAME 1 is basic	NAME 2 is nonbasic at its lower bound
XU	NAME 1 is basic	NAME 2 is nonbasic at its upper bound (applies only to ranged rows).
LL	NAME 1 is nonbasic at its lower bound	
UL	NAME 1 is nonbasic at its upper bound (applies only to bounded variables).	

Field 3 should be blank (and is ignored) for the LL and UL basis status.

GETOFF: Producing an advanced basis The same file discussed in the previous section can be produced on termination of the current run. It is called the

PUNCH file in the MPS series of codes, and the BASISOUT file in the APEX series of codes.

This file can be used directly when restarting on the same problem, or possibly a similar problem with a different cost row (a different right-hand side vector may cause infeasibilities). It is possible to modify the file for subsequent runs, although it is only rarely that one would wish to do this: for instance, when adding or deleting variables.

9-6 EXTENSIONS

This section will discuss how data are organized for commercial systems which have the extended facilities of generalized upper bounding, parametric programming, separable programming, and mixed-integer programming. Use of these facilities requires little additional data preparation to what has already been discussed in this chapter; the other extensions discussed in Part I require specialized data preparation and will not be discussed here.

9-6-1 Generalized Upper Bounding

GUB rows are *not* declared in the ROWS section of the MPS-format data input file. Instead, variables which are members of GUB sets are indicated in the COLUMNS section. *Marker* cards are used to indicate the start and end of the columns belonging to a particular GUB set. They have the format as shown below:

	Field 1 (2–3)	Field 2 (5–12)	Field 3 (15–22)	Field 4 (25–36)	Field 5 (40–47)	Field 6 (50–61)
Start marker	G, L, N, or E	(GUB rowlabel)	'MARKER'	blank	'GUBORG'	blank
Columns data cards		(Col. label)	(Row label)	(Value)	(Row label)	(Value)
End marker	blank	(Marker name)	'MARKER'	blank	'GUBEND'	blank

The start marker card specifies the GUB row type in field 1, the GUB row label in field 2, and has 'MARKER' and 'GUBORG' (including the quotation marks) in fields 3 and 5 respectively.

The columns data cards are exactly the same as those discussed in Sec. 9-2. They specify the nonzero row entries in both the ordinary rows and the GUB row (for which a default value of $+1.0$ is assumed if none is specified) in each of the columns belonging to the GUB set.

The end marker card has a marker name in field 2 (any unique name will do), as well as 'MARKER' and 'GUBEND' in fields 3 and 5 respectively.

The right-hand side of each GUB row is specified in the RHS section as if it was an ordinary row, placing the GUB row label in field 3. Similarly, each GUB row can be ranged as if it was an ordinary row by making the appropriate entries in the RANGES section.

Example Consider the following problem:

$$
\begin{array}{llcccccl}
\text{Maximize} & x_1 & + x_2 & + 2x_3 & + 3x_4 & + x_5 & & \\
\text{subject to} & x_1 & & + 3x_3 & & + x_5 & \leqslant & 8 \\
& & x_2 & + 2x_3 & + \ x_4 & & \leqslant & 10 \\
& & x_2 & + \ x_3 & & & \leqslant & 6 \\
& & & & x_4 & + x_5 & = & 2
\end{array}
$$

Specifying the last two rows as GUB rows gives rise to the MPS file shown in Table 9-3.

Table 9-3 MPS file for GUB example

```
NAME          GUB PROB
ROWS
 L  ROW1
 L  ROW2
 N  PROFIT
COLUMNS
    X1        ROW1       1.0        PROFIT     1.0
 L  GUB1      'MARKER'              'GUBORG'
    X2        ROW2       1.0        PROFIT     1.0
    X3        ROW1       3.0        ROW2       2.0
    X3        PROFIT     2.0
    SET1      'MARKER'              'GUBEND'
 E  GUB2      'MARKER'              'GUBORG'
    X4        ROW2       1.0        PROFIT     3.0
    X5        ROW1       1.0        PROFIT     1.0
    SET2      'MARKER'              'GUBEND'
RHS
    B         ROW1       8.0        ROW2       10.0
    B         GUB1       6.0        GUB2       2.0
ENDATA
```

9-6-2 Parametric Programming

Parametric programming facilities can be called by using control commands. The only extra information which is required is the change row $\hat{\mathbf{c}}$ specified as a free row (N) in the MPS input file for parametric variation of the cost row, or the change column $\hat{\mathbf{b}}$ specified as another right-hand side (with its own label) in the RHS section for parametric variation of the right-hand side.

The information to be specified is as follows:

1. *Parametric variation of cost row*
   Change cost row label ($\hat{\mathbf{c}}$) (it must be declared as type N in the ROWS section).
   Value of $\theta_{\max}$.
   Command to vary cost row (e.g., 'PARAOBJ').

2. *Parametric variation of RHS*
   Change RHS label ($\hat{\mathbf{b}}$) (it must be in RHS section, not COLUMNS).
   Value of θ_{max}.
   Command to vary right-hand side (e.g., 'PARARHS').

An iteration log is usually printed at every change of basis as θ increases; however, it is possible to specify the increments of θ at which a solution is printed.

9-6-3 Separable Programming

The separable-programming procedure discussed in Sec. 6-1 is the *lambda method*, and it turns out that generalized upper bounding can be used to great effect in this procedure.

In particular, the *convexity* rows for special variables (cf. Eq. 6-7) are of the GUB variety. The special variables are thus designated as GUB sets, with the following special features:

1. The right-hand side is always 1.0.
2. The row type declared in the marker card is always E (equality).
3. The special variables have ordinary bounds ($\geqslant$0.0), so do not have any entries in the BOUNDS section.
4. The marker cards have 'GSPORG' instead of 'GUBORG', and 'GSPEND' instead of 'GUBEND'.

The other separable-programming rows (Eqs. (6-8), (6-9)) become part of the ordinary rows of the problem. The special markers indicate to the system that no more than two contiguous special variables from each set can be basic (nonzero) simultaneously.

9-6-4 Mixed-integer Programming

A *pure* integer program can be denoted by an appropriate command statement in the agenda cards, but a *mixed* integer program requires the user to specify which of the variables are required to be integer.

Marker cards can be used to specify the start and end of a group of integer variables in the same manner as GUB sets, discussed in Sec. 9-6-1. The start marker has the marker name in field 2, the keyword 'MARKER' in field 3, and the keyword 'INTORG' in field 5. The end marker has the marker name in field 2, the keyword 'MARKER' in field 3, and the keyword 'INTEND' in field 5.

9-6-5 Special Ordered Sets

Special ordered sets are specified in the BOUNDS section. All the variables belonging to a set must appear contiguously in the COLUMNS section, so that one card may be used to specify the start and end of the set.

We thus have one card in the BOUNDS section for each special ordered set. Field 1 contains the keyword S1 or S2, for Type 1 or Type 2 set respectively. Field 2 contains the bounds row label, which will be the same as that for all the other bounded variables previously defined in the BOUNDS section. Field 3 contains the column label of the first variable in the special ordered set. Field 5 contains the column label of the last variable in the set.

Note that no variable in a special ordered set may have a bound on it, nor may it be declared as integer or a member of another special ordered set. Type 1 sets must contain at least two variables, and Type 2 sets must contain at least three variables.

CHAPTER

TEN

COMMERCIAL SYSTEMS: INTERPRETATION OF OUTPUT

10-1 INTRODUCTION

This chapter discusses the output produced from commercial codes, and how the data may be used in the analysis of the problem under consideration. While all commercial codes differ in the output they produce, it is possible to make some general remarks about the information that is printed. They all have the option to print out the constraints row by row for error checking, although the format for this differs. They also, as a matter of course, print out the information entered on the control commands (Sec. 9-3) and the matrix statistics: number of rows (by type), number of columns (by type), number of nonzeros, matrix density, and the average nonzeros per row and column.

An iteration log can be requested, and this gives such information as the number of pivots between each reinversion, processing time, utilization of storage, number of infeasibilities (during Phase I), and the number of nonoptimal nonbasic variables (i.e., those whose reduced costs are of the wrong sign for optimality).

Fortunately, the most important information, the current solution at the termination of a run, is presented in the same way by all codes, and this is discussed in the next section.

10-2 MPS OUTPUT FORMAT

The output listing produced on termination of a run has become standardized throughout the industry. There are a few minor variations (for example, some codes produce additional information, such as row type for the rows and bound type for the variables), but the presentation and layout are the same.

It is assumed that the output is printed, but it is usually possible to copy it to another file for further processing. It is also possible to specify that only a portion of the output is printed, for example only the basic variables.

The output is headed by information about the problem, and then followed by two sections: a ROWS section and a COLUMNS section.

The header information is as follows:

Problem name. The label given in the NAME section of the MPS-formatted input data (Sec. 9-2).
Objective value. The value of the objective function P.
Status. Optimal, nonoptimal, or infeasible.
Iterations. The total number of pivot steps taken so far.
Objective. Row-label of cost row.
RHS. Column-label of RHS column.
Ranges. Column-label of RANGES column.
Bounds. Row-label of BOUNDS row.

ROWS section Data are specified for each row under the following headings:

1. NUMBER – row number according to input order.
2. NAME – row label.
3. AT (STATUS) – status of the row activity; if a constraint is of the form

$$h_i \leqslant \sum_{j=1}^{n} a_{ij}x_j \leqslant u_i$$

the row activity is defined as $\sum_{j=1}^{n} a_{ij}x_j$, and one of the following codes is given for the value of this row activity:

(BINDING) { LL – at lower limit h_i
UL – at upper limit u_i
EQ – fixed ($h_i = u_i$, i.e., equality constraint)

(SLACK) BS – basic (between the limits)
(i.e., slack variable is basic)

Asterisks indicate a degeneracy.

4. ACTIVITY – value of the row activity, $\sum_{j=1}^{n} a_{ij}x_j$.
5. SLACK ACTIVITY – value of the slack variable. The values of items 4 and 5 should add to either h_i or u_i, except for a free row when they should sum to zero.
6. LOWER LIMIT (RHS LOWER) – value of h_i (-1.0×10^{30} or $-$INF or NONE if no lower bound, i.e., for an L ($\leqslant$) row type with no range).
7. UPPER LIMIT (RHS UPPER) – value of u_i ($+1.0 \times 10^{30}$ or $+$INF or NONE if no upper bound, i.e., for a G($\geqslant$) row type with no range).
8. DUAL ACTIVITY (MARGINAL) – the value of the shadow price for that constraint. (Recall from Secs. 4-1 and 4-2 that the values of the dual variables $\mathbf{y}$ are the reduced costs of the primal slack variables, and also correspond to the pricing vector π at the optimum.)

Although the same information is printed, the terminology used varies between codes; for example, the terms in brackets above are used by the APEX series of codes, while the other terms are used by the MPS series of codes.

COLUMNS section Data are specified for each column (variable) under the following headings:

1. NUMBER – column number according to input order (usually incremented by m, the number of rows).
2. NAME – column label.
3. AT (STATUS) – status of the variable. Asterisks indicate a degeneracy. The code is one of the following:
   (LOWER) LL – nonbasic, at lower limit
   (UPPER) UL – nonbasic, at upper limit
   (ACTIVE) BS – basic (feasible)
   FR – nonbasic, free
   EQ – nonbasic, fixed
   XX – basic, infeasible
4. ACTIVITY – the value of the variable (x_j).
5. INPUT COST (OBJCOEF) – the corresponding cost coefficient (c_j).
6. LOWER LIMIT (BND LOWER) – the lower bound on the variable (0.0 unless redefined in the BOUNDS section).
7. UPPER LIMIT (BND UPPER) – the upper bound on the variable (+1.0E30 unless redefined in the BOUNDS section).
8. REDUCED COST (MARGINAL) – the reduced cost of the variable (d_j). (*NOTE:* according to the convention we have used, the reduced cost should be nonnegative ($\geqslant 0$) for a nonbasic variable at its lower bound, and nonpositive ($\leqslant 0$) for a nonbasic variable at its upper bound. However, the convention differs between codes, so the reverse may apply; by checking the status (item 3 above) it will be evident what the appropriate sign is. In APEX, the reverse convention applies, as will be seen in the example output; i.e. the reduced cost is defined as $c_j - z_j$).

Example The output for our example problem is shown in Table 10-1.

10-3 RANGING PROCEDURES

We have already discussed the theory of ranging in Sec. 4-4, so we will limit our discussion in this section to the type of output produced when the RANGE option is specified.

Accompanying this option are further commands which may be used to restrict the analyses to certain specific rows and columns, for example only the rows whose slack variables are basic and only the columns corresponding to basic variables. Since ranging is expensive (in time and output volume), it is usual to specify only a portion of the entire analysis.

Table 10-1 MPS output file for example problem

C O N S T R A I N T S PAGE 1

PRINT OPTION = COMPLETE OUTPUT VALUE OF OBJECTIVE = 126.05712
NAME = OILREFINE OBJ = PROFIT RHS = LIMTMAX BND = RPSOBJ = -1.0000 RPSRHS = 1.0000
DIR = MAXIMIZE COBJ = CRHS = RNG = RPCHOBJ = .0000 RPCHRHS = .0000

NUMBER	NAME	TYPE		STATUS	ROW ACTIVITY	SLACK	RHS LOWER	RHS UPPER	MARGINAL
1	PROFIT	FR		SLACK	126.05712	-126.05712	-INF	+INF	.
2	MVOLBOL	LE		BINDING	26.31600	.	-INF	26.31600	-.97667
3	MVOLCOL	LE		SLACK	6.70641	14.34559	-INF	21.05200	.
4	MVOLLNC	EQ		BINDING	.	.	.	.	-16.52757
5	MVOLLNB	EQ		BINDING	.	.	.	.	-16.03421
6	MVOLSRK	EQ		BINDING	.	.	.	.	-14.40000
7	MVOLSRD	EQ		BINDING	.	.	.	.	-14.40000
8	MVOLVBB	EQ		BINDING	.	.	.	.	-8.00000
9	MVOLVBC	EQ		BINDING	.	.	.	.	-5.41958
10	MVOLRCR	EQ		BINDING	.	.	.	.	-14.79775
11	MVOLHVO	EQ		BINDING	.	.	.	.	-5.60185
12	UBALKWH	EQ		BINDING	.	.	.	.	-.04000
13	UBALH2O	EQ		BINDING	.	.	.	.	-.05280
14	UBALSTM	EQ		BINDING	.	.	.	.	-.16000
15	UBALFUL	EQ	**	BINDING	.	.	.	.	.
16	MVOLB95	EQ		BINDING	.	.	.	.	-9.86953
17	MVOLB90	EQ		BINDING	.	.	.	.	-9.92896
18	MVOLLHG	EQ		BINDING	.	.	.	.	-10.35173
19	MVOLC3S	EQ		BINDING	.	.	.	.	-11.00000
20	MVOLNC4	EQ		BINDING	.	.	.	.	-11.00000
21	MVOLLSR	EQ		BINDING	.	.	.	.	-17.30882
22	MVOLHSR	EQ		BINDING	.	.	.	.	-16.74912
23	MVOLIC4	EQ		BINDING	.	.	.	.	-11.00000
24	VCAPSGP	LE		SLACK	16.09734	7.16266	-INF	23.26000	.
25	MVOLRFG	EQ		BINDING	.	.	.	.	-16.80081
26	MSCFHYL	EQ	**	BINDING	.	.	.	.	.
27	MVOLR90	EQ		BINDING	.	.	.	.	-16.80081
28	MVOLR95	EQ		BINDING	.	.	.	.	-16.77532
29	MVOLF90	EQ		BINDING	.	.	.	.	-19.94606
30	MVOLF95	EQ		BINDING	.	.	.	.	-21.10403
31	VCAPRFG	LE		BINDING	13.45500	.	-INF	13.45500	-.48255
32	MVOLLCO	EQ		BINDING	.	.	.	.	-13.81360
33	MVOLHHG	EQ		BINDING	.	.	.	.	-16.80081
34	MVOLHCO	EQ		BINDING	.	.	.	.	-18.96617
35	VCAPHVO	LE		BINDING	5.25000	.	-INF	5.25000	-9.19589
36	VCAPHOL	LE		BINDING	3.87000	.	-INF	3.87000	-4.53708
37	MVOLC3U	EQ		BINDING	.	.	.	.	-23.54955
38	MVOLC4U	EQ		BINDING	.	.	.	.	-28.72026
39	MVOLFCG	EQ		BINDING	.	.	.	.	-19.72558
40	MVOLSLR	EQ		BINDING	.	.	.	.	-10.09086
41	VCAPCCU	LE		BINDING	7.26000	.	-INF	7.26000	-3.86024
42	MVOLLA3	EQ		BINDING	.	.	.	.	-22.58708
43	MVOLLA4	EQ		BINDING	.	.	.	.	-23.59616
44	VCAPALK	LE		SLACK	1.84763	8.15237	-INF	10.00000	.
45	XLPRPRE	LE		SLACK	-5.19322	5.19322	-INF	.	.
46	XHPRPRE	LE		BINDING	.	.	-INF	.	-4.63189
47	XTELPRE	LE		BINDING	.	.	-INF	.	-.39466
48	XRVPPRE	LE		BINDING	.	.	-INF	.	-.18612
49	X200PRF	LE		SLACK	-.88476	.88476	-INF	.	.
50	X230PRF	LE		SLACK	-.37305	.37305	-INF	.	.

```
PRINT OPTION = COMPLETE OUTPUT                                          VALUE OF OBJECTIVE =           126.05712
NAME = OILREFINE   OBJ  = PROFIT    RHS  = LIMTMAX    BND =             RPSOBJ  =     -1.0000  RPSRHS  =     1.0000
DIR  = MAXIMIZE    COBJ =           CRHS =            RNG =             RPCHOPJ =       .0000  RPCHRHS =      .0000
```

NUMBER	NAME	TYPE	STATUS	ROW ACTIVITY	SLACK	RHS LOWER	RHS UPPER	MARGINAL
51	EVOLPRE	EQ	BINDING	.	.	.	.	23.75575
52	XPSCPRE	LE	SLACK	-.04312	.04312	-INF	.	.
53	XRSCREG	LE	BINDING	.	.	-INF	.	-1.44575
54	XLPRINT	LE	SLACK	-2.17940	2.17940	-INF	.	.
55	XHPRINT	LE	BINDING	.	.	-INF	.	-1.77692
56	XTELINT	LE	SLACK	-4.13033	4.13033	-INF	.	.
57	XRVPINT	LE	BINDING	.	.	-INF	.	-.16233
58	X200INT	LE	SLACK	-.63517	.63517	-INF	.	.
59	X230INT	LE	BINDING	.	.	-INF	.	-.25325
60	EVOLINT	EQ	BINDING	.	.	.	.	-4.08218
61	XLPRREG	LE	BINDING	.	.	-INF	.	-.86761
62	XHPRREG	LE	SLACK	-.44308	.44308	-INF	.	.
63	XTELREG	LE	SLACK	-20.71386	20.71386	-INF	.	.
64	XRVPREG	LE	BINDING	.	.	-INF	.	-.15119
65	X200REG	LE	BINDING	.	.	-INF	.	-.14180
66	X230REG	LE	SLACK	-.92614	.92614	-INF	.	.
67	EVOLREG	EQ	BINDING	.	.	.	.	-12.63510
68	EVOLLPG	EQ	BINDING	.	.	.	.	-11.00000
69	EVOLJP4	EQ	BINDING	.	.	.	.	-16.71520
70	XRVXJP4	LE	SLACK	-.23921	.23921	-INF	.	.
71	XRVNJP4	LE	BINDING	.	.	-INF	.	-.04240
72	EVOLDSL	EQ	BINDING	.	.	.	.	-14.40000
73	EVOLRSD	EQ	BINDING	.	.	.	.	-18.30129
74	XVISRSD	LE	BINDING	.	.	-INF	.	-1.01993

Table 10-1 continued.

C O L U M N S   PAGE 1

```
PRINT OPTION = COMPLETE OUTPUT                                     VALUE OF OBJECTIVE =                    126.05712
NAME = OILREFINE   OBJ  = PROFIT      RHS  = LIMTMAX     BND =     RPSOBJ  =     -1.0000   RPSRHS  =    1.0000
DIR  = MAXIMIZE    COBJ =             CRHS =             RNG =     RPCHOBJ =      .0000    RPCHRHS =     .0000
```

NUMBER	NAME	TYPE		STATUS	COL ACTIVITY	OBJ COFF	BND LOWER	BND UPPER	MARGINAL
1	VCROBOL	PL		ACTIVE	26.31600	-12.80000	.	+INF	.
2	VCRDCOL	PL		ACTIVE	6.70641	-11.48000	.	+INF	.
3	VSGPLNC	PL		ACTIVE	1.96565	.	.	+INF	.
4	VSGPLNB	PL		ACTIVE	14.13169	.	.	+INF	.
5	VSGPLHG	PL		ACTIVE	1.13706	.	.	+INF	.
6	VSGPB95	PL		ACTIVE	.	.	.	+INF	.
7	VSGPB90	PL		ACTIVE	.54358	.	.	+INF	.
8	VH2RHSR	PL		ACTIVE	11.05853	-.01750	.	+INF	.
9	VRFFRF1	PL		ACTIVE	11.05853	.	.	+INF	.
10	VRFFRF2	PL		LOWER	.	.	.	+INF	-.02550
11	VRFFHH1	PL		ACTIVE	2.39647	.	.	+INF	.
12	VRFFHH2	PL		LOWER	.	.	.	+INF	-.02550
13	VRFGR90	PL		ACTIVE	13.45500	-.15120	.	+INF	.
14	VRFGR95	PL		ACTIVE	.	-.30400	.	+INF	.
15	VHOLLCO	PL		ACTIVE	1.63362	-.21120	.	+INF	.
16	VHOLSRD	PL		LOWER	.	-.21120	.	+INF	-.58640
17	VHOLRCR	PL		ACTIVE	1.78743	-.21120	.	+INF	.
18	VHOLHVO	PL		ACTIVE	.44835	-.21120	.	+INF	.
19	VCCUSRK	PL		ACTIVE	2.45895	-.21120	.	+INF	.
20	VCCUSRD	PL	**	LOWER	.	-.21120	.	+INF	.
21	VCCURCR	PL	**	LOWER	.	-.21120	.	+INF	.
22	VCCUHVO	PL		ACTIVE	4.80105	-.21120	.	+INF	.
23	VALKLA3	PL		ACTIVE	.83474	-.51200	.	+INF	.
24	VALKLA4	PL		ACTIVE	1.01239	-.47200	.	+INF	.
25	VALKIC4	PL		ACTIVE	.03285	.	.	+INF	.
26	VALKC3U	PL		LOWER	.	.	.	+INF	-12.54955
27	VALKC4U	PL		LOWER	.	.	.	+INF	-17.72026
28	UTILC3S	PL		LOWER	.	.	.	+INF	-11.00000
29	UTILNC4	PL		LOWER	.	.	.	+INF	-11.00000
30	UTILIC4	PL		LOWER	.	.	.	+INF	-11.00000
31	UTILC3U	PL		LOWER	.	.	.	+INF	-23.54955
32	UTILC4U	PL		LOWER	.	.	.	+INF	-28.72026
33	UTILHYL	PL		ACTIVE	1.82296	.	.	+INF	.
34	UTILSTM	PL		ACTIVE	.27078	-.16000	.	+INF	.
35	PURCPC4	PL		LOWER	.	-12.00000	.	+INF	-1.00000
36	PURCH2O	PL		ACTIVE	25.09950	-.05280	.	+INF	.
37	PURCKWH	PL		ACTIVE	105.56916	-.04000	.	+INF	.
38	PURCFUL	PL		LOWER	.	-1.60000	.	+INF	-1.60000
39	PURCFLR	PL		ACTIVE	.35327	.	.	+INF	.
40	BLPGC3S	PL		ACTIVE	1.04958	.	.	+INF	.
41	BLPGNC4	PL		ACTIVE	.18324	.	.	+INF	.
42	SELLLPG	PL		ACTIVE	1.23282	11.00000	.	+INF	.
43	BJP4LSR	PL		ACTIVE	.02175	.	.	+INF	.
44	BJP4HSP	PL		ACTIVE	.21747	.	.	+INF	.
45	SELLJP4	PL		ACTIVE	.23921	16.80000	.	+INF	.
46	BDSLSRK	PL		ACTIVE	1.77310	.	.	+INF	.
47	BDSLSRD	PL		ACTIVE	3.47474	.	.	+INF	.
48	SELLDSL	PL		ACTIVE	5.24734	14.40000	.	+INF	.
49	BPRELSR	PL		LOWER	.	.	.	+INF	-2.18881
50	BPREHCO	PL		ACTIVE	1.01202	.	.	+INF	.

```
PRINT OPTION = COMPLETE OUTPUT                                              VALUE OF OBJECTIVE =         126.05712
NAME = OILREFINE    OBJ  = PROFIT        RHS  = LIMTMAX      BND =          RRSOBJ  =     -1.0000   RPSRHS  =    1.0000
DIR  = MAXIMIZE     COBJ =               CRHS =              RNG =          RRCHOBJ =       .0000   RPCHRHS =     .0000
```

NUMBER	NAME	TYPE	STATUS	COL ACTIVITY	OBJ COEF	BND LOWER	BND UPPER	MARGINAL
51	BPREF95	PL	ACTIVE	.	.	.	+INF	.
52	BPREF90	PL	ACTIVE	4.60270	.	.	+INF	.
53	BPREFCG	PL	LOWER	.	.	.	+INF	-.70802
54	BPRELA3	PL	ACTIVE	.83474	.	.	+INF	.
55	BPRELA4	PL	ACTIVE	1.01289	.	.	+INF	.
56	BPRENC4	PL	ACTIVE	.65857	.	.	+INF	.
57	BPRETEL	PL	ACTIVE	24.36273	-.36960	.	+INF	.
58	SELLPRE	PL	ACTIVE	8.12091	21.44000	.	+INF	.
59	BINTLSR	PL	ACTIVE	.94144	.	.	+INF	.
60	BINTHCD	PL	ACTIVE	.27620	.	.	+INF	.
61	BINTF95	PL	LOWER	.	.	.	+INF	-.70741
62	BINTF90	PL	ACTIVE	4.96230	.	.	+INF	.
63	BINTFCG	PL	LOWER	.	.	.	+INF	-.26516
64	BINTNC4	PL	ACTIVE	.43977	.	.	+INF	.
65	BINTTEL	PL	ACTIVE	15.72880	-.36960	.	+INF	.
66	SELLINT	PL	ACTIVE	6.61971	20.32000	.	+INF	.
67	BREGLSR	PL	ACTIVE	2.18996	.	.	+INF	.
68	BREGHCD	PL	LOWER	.	.	.	+INF	-.58280
69	BREGF95	PL	LOWER	.	.	.	+INF	-.81490
70	BREGF90	PL	ACTIVE	1.95787	.	.	+INF	.
71	BREGFCG	PL	ACTIVE	3.53684	.	.	+INF	.
72	BREGNC4	PL	ACTIVE	.25259	.	.	+INF	.
73	BREGTEL	PL	ACTIVE	3.09790	-.36960	.	+INF	.
74	SELLREG	PL	ACTIVE	7.93726	18.04000	.	+INF	.
75	BRSDVBB	PL	ACTIVE	.97369	.	.	+INF	.
76	BRSDVBC	PL	ACTIVE	1.20715	.	.	+INF	.
77	BRSDRCR	PL	LOWER	.	.	.	+INF	-3.53397
78	BRSDHVO	PL	LOWER	.	.	.	+INF	-4.70689
79	BRSDSLR	PL	ACTIVE	.04609	.	.	+INF	.
80	BRSDLCO	PL	ACTIVE	.51923	.	.	+INF	.
81	SELLRSD	PL	ACTIVE	2.74617	8.00000	.	+INF	.

The output is divided into two sections, ROWS and COLUMNS. Note that the column label of a slack variable is the same as the row label of its corresponding row.

ROWS section Data are printed on two lines for each row under the following headings:

1. NUMBER – row number.
2. NAME – row label.
3. AT (STATUS) – status of the row activity, $\sum_{j=1}^{n} a_{ij}x_j$.

   (BINDING) { LL – at lower limit h_i
   UL – at upper limit u_i
   EQ – fixed ($h_i = u_i$)

   (SLACK) BS – basic slack variable ($\sum a_{ij}x_j$ is between the limits)
4. ACTIVITY – value of $\sum_{j=1}^{n} a_{ij}x_j$.
5. SLACK ACTIVITY – value of the slack variable.
6. (MARGINAL) – the value of the shadow price for the constraint.
7. LOWER LIMIT (RHS LOWER) – value of h_i.
8. UPPER LIMIT (RHS UPPER) – value of u_i.

   (The above data are analogous to those discussed in Sec. 10-2.)

9. LOWER ACTIVITY – the minimum value the row activity can attain, regardless of h_i, before a basis change is required. The cost per unit of decrease is given by UNIT COST. (If the UNIT COST is positive, the objective function gets worse, while if it is negative, the objective function improves.)
10. UPPER ACTIVITY – the maximum value the row activity can attain, regardless of u_i, before a basis change is required. The cost per unit of increase is given by UNIT COST.
11. (OBJ LOWER)
    (OBJ UPPER) – the values of the objective function corresponding to the row activity being at LOWER ACTIVITY and UPPER ACTIVITY respectively. (The objective function changes by UNIT COST multiplied by the amount of decrease or increase.)
12. LIMITING PROCESS – the name of the column or row (slack variable) which would change its status on further movement of the row activity beyond LOWER ACTIVITY or UPPER ACTIVITY. For a row whose STATUS is BINDING, the named variable will leave the basis, while for a row whose STATUS is SLACK, the named variable will enter the basis.
13. UPPER COST (OBJ COEF)
    LOWER COST (RANGE) – the range over which the cost coefficient on the slack variable (which is usually zero) may vary without a change in the current basis. For a row whose STATUS is SLACK, the variable named under LIMITING PROCESS is the nonbasic variable whose reduced cost goes to zero (so it may enter the basis).

14. (OBJ at OBJ COEF RANGE) – the corresponding value of the objective function. (The objective changes by OBJ COEF RANGE multiplied by SLACK ACTIVITY.)

COLUMNS section Data are printed on two lines for each column under the following headings:

1. NUMBER – column number.
2. NAME – column label.
3. AT (STATUS) – status of the variable:
   (LOWER) LL – nonbasic, at lower limit
   (UPPER) UL – nonbasic, at upper limit
   (ACTIVE) BS – basic
   EQ – nonbasic, fixed
   FR – nonbasic, free
4. ACTIVITY – the value of the variable (x_j).
5. INPUT COST (OBJ COEF) – the corresponding cost coefficient (c_j).
6. (MARGINAL) – the reduced cost of the variable (d_j).
7. LOWER LIMIT (BND LOWER) – lower bound on the variable.
8. UPPER LIMIT (BND UPPER) – upper bound on the variable.
9. LOWER ACTIVITY – the minimum value that the variable can attain, regardless of the specified lower bound, before a basis change is required. The cost per unit of decrease is given by UNIT COST.
10. UPPER ACTIVITY – the maximum value that the variable can attain, regardless of the specified upper bound, before a basis change is required. The cost per unit of increase is given by UNIT COST.
11. (OBJ LOWER)
    (OBJ UPPER) – the values of the objective function corresponding to the variable being at LOWER ACTIVITY and UPPER ACTIVITY respectively. (The objective function changes by UNIT COST multiplied by the amount of decrease or increase.)
12. LIMITING PROCESS – the name of the column or row (slack variable) which would change its status on further movement of the variable beyond LOWER ACTIVITY or UPPER ACTIVITY. For a variable whose STATUS is ACTIVE, i.e., a basic variable, the named variable will enter the basis, while for a variable whose STATUS is UPPER or LOWER, i.e., a nonbasic variable, the named variable will leave the basis.
13. UPPER COST (OBJ COEF)
    LOWER COST (RANGE) – the range over which the cost coefficient may vary without a change in the current basis. This is the range of $\bar{c}_j$, *not* the range of δ (cf. Sec. 4-4). For a variable whose STATUS is ACTIVE, the variable named under LIMITING PROCESS is the nonbasic variable whose reduced cost goes to zero (so it may enter the basis).
14. (OBJ at OBJ COEF RANGE) – the corresponding value of the objective function. (The objective changes by the difference (δ) between $\bar{c}_j$ and INPUT COST (c_j), multiplied by ACTIVITY.)

Table 10-2

```
NAME           ORE PROCESSING
ROWS
 L  R1(X9)
 G  R2(X10)
 L  R3(X8)
 L  R4(X11)
 L  R5(X12)
 L  R6(X13)
 L  R7(X14)
 N  PROFIT
COLUMNS
    X1          R1(X9)      0.15          R2(X10)    0.15
    X1          R5(X12)     1.0           R7(X14)    -0.85
    X1          PROFIT      0.44
    X2          R1(X9)      0.25          R2(X10)    0.25
    X2          R3(X8)      1.0           R5(X12)    1.0
    X2          R7(X14)     -0.75         PROFIT     0.45
    X3          R1(X9)      1.0           R2(X10)    1.0
    X3          PROFIT      -0.25
    X4          R7(X14)     -1.0          PROFIT     -0.2
    X5          R1(X9)      0.5           R2(X10)    0.5
    X5          R6(X13)     1.0           R7(X14)    1.0
    X5          PROFIT      0.6
    X6          R1(X9)      -1.0          R2(X10)    -1.0
    X6          R4(X11)     1.0           PROFIT     -0.3
    X7          R1(X9)      -1.0          R2(X10)    -1.0
    X7          PROFIT      -0.5
RHS
    LIM         R1(X9)      45.0          R2(X10)    40.0
    LIM         R3(X8)      30.0          R4(X11)    4.0
    LIM         R5(X12)     100.0         R6(X13)    50.0
ENDATA
```

Example To see how all this information relates to what has been discussed in Chapter 4 it will be instructive to consider a RANGE output for the example of Sec. 4-6. The MPS-format input data are shown in Table 10-2.

Note that by expressing Product I normal sales at 5.50 \$/tonne in terms of the other variables,

$$\text{Normal sales} = 0.15x_1 + 0.25x_2 + x_3 + 0.5x_5 - x_6 - x_7$$

the objective row becomes:

$$\text{Maximize } P = 0.44x_1 + 0.45x_2 - 0.25x_3 - 0.2x_4 + 0.6x_5 - 0.3x_6 - 0.5x_7$$

Thus, for example, the objective coefficient for x_1 in the model is $+0.44$, *not* -3.25, and for x_4 it is -0.2, *not* -4.0. The reader will find it useful to maintain reference to Sec. 4-6, especially the final tableau, in going through this analysis.

By considering the RHS section of data it will be seen that the slack variables in Sec. 4-6 correspond to the following rows:

Row	Slack variable (Sec. 4-6)
R1(X9)	x_9
R2(X10)	x_{10}
R3(X8)	x_8
R4(X11)	x_{11}
R5(X12)	x_{12}
R6(X13)	x_{13}
R7(X14)	x_{14}

The output produced by the RANGE option is shown in Table 10-3.

Table 10-3

C O N S T R A I N T S

PAGE 1

PRINT OPTION = COMPLETE OUTPUT
NAME = OREPROCES OBJ = PROFIT RHS = LIM BND = VALUE OF OBJECTIVE = 74.30000
DIR = MAXIMIZE COBJ = CRHS = RNG = RPSOBJ = -1.0000 RPSRHS = 1.0000
RPCHOBJ = .0000 RPCHRHS = .0000

NUMBER	NAME TYPE	ROW ACTIVITY STATUS	SLACK MARGINAL	RHS LOWER RHS UPPER	LOWER ACT UPPER ACT	UNIT COST UNIT COST	OBJ&LOWER OBJ&UPPER	LIMITING PROCESS	OBJ COEF RANGE	OBJ&OBJ COEF RANGE
1	R1(X9) LE	43.00000 SLACK	2.00000 .	-INF 45.00000	40.00000 +INF	.10000 .25000	74.30000 -INF	R3(X8) X3	.10000 -.25000	74.50000 73.80000
2	R2(X10) GE	43.00000 SLACK	-3.00000 .	40.00000 +INF	40.00000 45.00000	.10000 .25000	74.00000 73.80000	R3(X8) X3	.10000 -.25000	74.00000 75.05000
3	R3(X8) LE	30.00000 BINDING	. -.01000	-INF 30.00000	. 50.00000	.01000 -.01000	74.00000 74.50000	R2(X10) R1(X9)	.01000 -INF	74.30000 74.30000
4	R4(X11) LE	. SLACK	4.00000 .	-INF 4.00000	. 3.00000	+INF .30000	74.30000 73.40000	NONE X6	+INF -.30000	+INF 73.10000
5	R5(X12) LE	100.00000 BINDING	. -.44000	-INF 100.00000	80.00000 113.33333	.44000 -.44000	65.50000 80.16667	R2(X10) R1(X9)	.44000 -INF	74.30000 74.30000
6	R6(X13) LE	50.00000 BINDING	. -.60000	-INF 50.00000	44.00000 54.00000	.60000 -.60000	70.70000 76.70000	R2(X10) R1(X9)	.60000 -INF	74.30000 74.30000
7	R7(X14) LE	-32.00000 SLACK	32.00000 .	-INF .	-35.00000 -15.00000	.10000 .51765	74.00000 65.50000	R3(X8) R5(X12)	.10000 -.51765	77.50000 57.73529
8	PROFIT FR	74.30000 SLACK	-74.30000 .	-INF +INF						

C O L U M N S

PAGE 1

PRINT OPTION = COMPLETE OUTPUT
NAME = OREPROCES OBJ = PROFIT RHS = LIM BND = VALUE OF OBJECTIVE = 74.30000
DIR = MAXIMIZE COBJ = CRHS = RNG = RPSOBJ = -1.0000 RPSRHS = 1.0000
RPCHOBJ = .0000 RPCHRHS = .0000

NUMBER	NAME TYPE	COL ACTIVITY STATUS	OBJ COEF MARGINAL	BND LOWER BND UPPER	LOWER ACT UPPER ACT	UNIT COST UNIT COST	OBJ&LOWER OBJ&UPPER	LIMITING PROCESS	OBJ COEF RANGE	OBJ&OBJ COEF RANGE
1	X1 PL	70.00000 ACTIVE	.44000 .	. +INF	50.00000 100.00000	.44000 .01000	65.50000 74.00000	R5(X12) R3(X8)	. .45000	43.50000 75.00000
2	X2 PL	30.00000 ACTIVE	.45000 .	. +INF	. 30.00000	.01000 +INF	74.00000 74.30000	R3(X8) NONE	.44000 +INF	74.00000 +INF
3	X3 PL	. LOWER	-.25000 -.25000	. +INF	-3.00000 2.00000	-.25000 .25000	75.05000 73.80000	R2(X10) R1(X9)	-INF .	74.30000 74.30000
4	X4 PL	. LOWER	-.20000 -.20000	. +INF	-32.00000 +INF	-.20000 .20000	80.70000 -INF	R7(X14) NONE	-INF .	74.30000 74.30000
5	X5 PL	50.00000 ACTIVE	.60000 .	. +INF	44.00000 50.00000	.60000 +INF	70.70000 74.30000	R6(X13) NONE	. +INF	44.30000 +INF
6	X6 PL	. LOWER	-.30000 -.30000	. +INF	-2.00000 3.00000	-.30000 .30000	74.90000 73.40000	R1(X9) R2(X10)	-INF .	74.30000 74.30000
7	X7 PL	. LOWER	-.50000 -.50000	. +INF	-2.00000 3.00000	-.50000 .50000	75.30000 72.80000	R1(X9) R2(X10)	-INF .	74.30000 74.30000

The COLUMNS section will be discussed first. The nonbasic variables (x_3, x_4, x_6, x_7) were discussed in Sec. 4-6-3(a): when the cost coefficient increases by δ given under the UNIT COST heading, the reduced cost of the *same* nonbasic variable is zero. We can therefore increase (decrease) the nonbasic variable until a basic variable (given under LIMITING PROCESS) reaches the end of its range and leaves the basis. UPPER ACT and LOWER ACT list the value of the variable after the increase and decrease respectively. For example, x_4 is zero and has a reduced cost of 0.2 (-0.2 under MARGINAL; the sign convention is different.)

The current solution remains optimal for c_j in the range

$$\bar{c}_j = c_j + \delta$$

where

$$-\infty < \delta \leqslant 0.2$$

(cf. Eq. (4-54))

i.e.,

$$-\text{INF} < \bar{c}_j \leqslant 0.0$$

which is what appears under the heading OBJ COEF RANGE. When $\delta = 0.2$, the reduced cost of x_4 is zero, so x_4 can increase by an unlimited amount, or decrease by 32 to -32, when x_{14}, the slack variable on R7 (LIMITING PROCESS), becomes zero. Note that we have ignored the implied lower bound of zero on x_4 itself. Thus -32 and $+\text{INF}$ are the LOWER and UPPER ACTIVITIES for x_4. In Sec. 4-6-3(a) we only considered increasing (not decreasing) the nonbasic variables at zero, but a glance at the appropriate columns will show the same numbers as appeared there, in particular:

Variable	$\hat{d}_j(-\infty < \delta \leqslant \hat{d}_j)$	UPPER ACT	Leaving basis variable (LIMITING PROCESS)
x_4	0.2	INF	NONE
x_6	0.3	3	x_{10}(R2)
x_7	0.5	3	x_{10}(R2)
x_3	0.25	2	x_9(R1)

The above analysis, and the analysis in Sec. 4-6-3(a), assume we have increased c_j by $\hat{d}_j$ so the new reduced cost of the nonbasic variable is zero. We could alternatively adopt the viewpoint of not changing c_j at all, so that $\hat{d}_j$ is the per unit cost of increasing the nonbasic variable at the same value of c_j (and thus appears under the UNIT COST heading); OBJ LOWER and OBJ UPPER thus give the value of the objective function at either end of the range in which the nonbasic variable can be changed.

The basic variables x_1, x_2 (and x_5) can be analysed in the same way as in Sec. 4-6-3(b). If the cost coefficient is increased (or decreased) by the amounts δ shown under UNIT COST, then the reduced costs of the nonbasic variables listed under LIMITING PROCESS become zero, in which case they can enter the basis up to a limit given by the standard ratio test (excluding bounds on the basic variable itself). LOWER ACT and UPPER ACT give the corresponding extreme values of the basic variable itself, and OBJ COEF RANGE gives the new values of c_j. For example, for x_1 the solution remains optimal in the range $c_1 + \delta$, where $-0.44 \leqslant \delta \leqslant 0.01$, i.e., $0.0 \leqslant \bar{c}_1 \leqslant 0.45$ (OBJ COEF RANGE) (c_1 is $+0.44$, not -3.25).

At $\delta = +0.01$, the reduced cost of x_8 is zero (LIMITING PROCESS (R3(X8)), and x_8 can increase up to a limit of 30, when $x_1 = 100.0$ (UPPER ACT). At $\delta = -0.44$, the reduced cost of x_{12} is zero (LIMITING PROCESS R5 (X12)), and x_{12} can increase up to a limit of 20, when $x_1 = 50.0$ (LOWER ACT). As before, we could adopt the alternative viewpoint of allowing the nonbasic variable indicated under LIMITING PROCESS to enter the basis, without changing the cost coefficient of the basic variable. This would incur the penalty, given by UNIT COST, per unit change in basic variable.

The ROWS section is analyzed in a similar manner. Note in particular if the STATUS is SLACK we can analyze the row entry in the same way as if we were analyzing the slack variable in the COLUMNS section and it was basic. For example, the slack variable on R1 (X9) is basic at 2.0; if its cost coefficient (0.0) changes by $\delta = +0.25$, the reduced cost of x_3 (LIMITING PROCESS) becomes zero, in which case x_3 can enter the basis in an unlimited amount if we ignore the bound on x_9 itself (check the tableau in Sec. 4-6). If the cost coefficient changes by $\delta = -0.1$, the reduced cost of x_8 (R3) becomes zero, in which case x_8 can increase up to 30.0. When x_8 is 30, x_9 is 5.0, (from the tableau), so the row activity is 40.0. Thus LOWER ACT and UPPER ACT are 40.0 and +INF respectively. (Note these are the *row* activities, *not* the activities of the slack variable).

Finally, the same analogy could be drawn between the row entry whose STATUS is BINDING and the corresponding slack variable analyzed as if it was a nonbasic variable in the COLUMNS section. Note, however, the shadow-price information discussed in Sec. 4-6-1 corresponds to the row entries under R5 (X12) and R6 (X13), and the price margin discussed in Sec. 4-6-2 corresponds to the row entry under R3 (X8):

R5 (X12): The shadow price on basic processing capacity is 0.44 \$/tonne in the range 80–113.3 M tonne/day.

R6 (X13): The shadow price on converter capacity is 0.6 \$/tonne in the range 44–54 M tonne/day.

R3 (X8): The price margin on type B ore is 0.01 \$/tonne in the range $X_2 = 0$–50 M tonne/day.

10-4 REPORT-WRITER LANGUAGES

While it can be acknowledged that the MPS output format discussed in Sec. 10-2 is a very compact way of presenting the optimal solution of the problem, it is hardly a suitable form of report to present for executive decision-making. Report-writer languages are a series of simple FORTRAN-like instructions to extract

Table 10-4 GAMMA instructions for report definition

```
 *
 *                         REPORT DEFINITION PHASE OF GAMMA
 *                         ********************************
 *
 *
        HEADER1,2
        HEADER2,2,T25='                          REFINERY COST MODEL'
        FORMAT,F(VOL)='XXX,XXX',F(CPC)='$XXXX,XXX',F(CPB)='$XXX.XX'
        FORMAT,F(PCT)='XXX.X',F(CST)='$XX,XXX.XX'
 *DO FOR EACH TABLE IN THE LIST (UNIT)
DO (UNIT)         ,FOR (ROWNAME)=((UNIT),,*),FOR(YYY)=((UNIT),*,)
                  ,FOR (XXXX)    =((UNIT),,*) $LIST OF ROWNAMESIN TABLE,4CHAR
                  ,IF((UNIT).NM.VRFF       UTIL PURC      )$EXCLUDE POOLS ETC.
PAGE
        SKIP4                                    $PRINT TEXT OF LIST(UNIT)
        LINE,T30=T(UNIT)
        LINE,T5='CHARGE STREAMS',T39='BBL/SDAY',T55='VOL PCT'
 *
        W(FT)=0.0                                $ARITHMETIC STATEMENT-NOTCLAUSE
        W(FT)=       CACT((UNIT)(YYY))*1000.+W(FT) $COL.ACTIVITES-TOTAL FEED
        W(PT)=0.0
        LINE (YYY)(ROWNAME) ,IF((XXXX/ROWNAME).IM.MVOL.AND.
                                   ((UNIT),(ROWNAME),(YYY)).EQ.1,0)
 *REPLICATE FOR EACH COL+ROW IF IT CORRESPONDS TO A FEED
             W(1)= CACT((UNIT)(YYY))*1000.0 $ARITHMETIC CLAUSE OF LINE STMT.
             W(2)= CACT((UNIT)(YYY))*1000.0/W(FT)
             W(PT)=W(2)*100.0+W(PT)    $PT IS TOTAL VOLPERCENT
             T8=((UNIT),(ROWNAME),T)   $PRINTS TEXT OF(ROWNAME)IN TABLE(UNIT)
             E(VOL)39=W(1)             $PRINTS W(1) IN FORMAT GRP(VOL)
             E(PCT)56=W(2)*100.0
        LINE,T38='-------',T55='-------'
        LINE,E(VOL)39=W(FT),E(PCT)56=W(PT),T11='TOTAL'
        SKIP 2
 *                    TABULATE   YIELD STREAMS
        LINE,T5='YIELD STREAMS'
        W(YT)=0.0
        W(YPCT)=0.0
        LINE (ROWNAME),SUPPRESS,IF((XXXX/ROWNAME).IM.MVOL) $DO YIELDS ONLY
             W(YT)=-((UNIT),(ROWNAME),(YYY))*CACT((UNIT)(YYY))*1000.+W(YT)
                  ,IF(((UNIT),(ROWNAME),(YYY)).LT.0.0)
 *   YIELD=COEFF*COL.ACTIVITY,NOTE THIS LINE CONTROL HAD NO PRINT INSERT
 *
        LINE (ROWNAME),SUPPRESS,IF((XXXX/ROWNAME).IM.MVOL)
             W(3)=0.
             W(3)=-((UNIT),(ROWNAME),(YYY))*CACT((UNIT)(YYY))*1000.+W(3)
                  ,IF(((UNIT),(ROWNAME),(YYY)).LT.0.0)
 *                             SUM OVER ALL CHARGE COLUMNS YYY
             W(4)=0.0
             W(4)=W(3)*100.0/W(YT),IF(W(YT).GT.0.0)   $CALC VOL PCT.
             W(YPCT)=W(4)+W(YPCT)
             T8=((UNIT),(ROWNAME),T)
             E(VOL)39=W(3)
             E(PCT)56=W(4)
        LINE,T39='-------',T55='-------'
        LINE,E(VOL)39=W(YT),E(PCT)56=W(YPCT) ,T11='TOTAL'
LOG,PAGE

        SKIP3
 *                  TABULATE UTILITY REQUIREMENTS
 *
        LINE,T5='UTILITY SUMMARY',T40='AMOUNT',T55='COST,$/SD'
        W(U1)=0.0   $ POWER
        W(U2)=0.0   $ COOLING WATER
        W(U3)=0.0   $ STEAM
        W(U4)=0.0   $ FUEL
        LINE (ROWNAME),SUPPRESS,IF((XXXX/ROWNAME).IM.UBAL)$DO UBAL ROWS ONLY
             W(U1)= ((UNIT),(ROWNAME),(YYY))*CACT((UNIT)(YYY))*1000.+W(U1)
                  ,IF((ROWNAME).IM.UBALKWH)
             W(U2)= ((UNIT),(ROWNAME),(YYY))*CACT((UNIT)(YYY))*1000.+W(U2)
                  ,IF((ROWNAME).IM.UBALH2O)
             W(U3)= ((UNIT),(ROWNAME),(YYY))*CACT((UNIT)(YYY))*1000.+W(U3)
                  ,IF((ROWNAME).IM.UBALSTM)
             W(U4)= ((UNIT),(ROWNAME),(YYY))*CACT((UNIT)(YYY))*1000.+W(U4)
                  ,IF((ROWNAME).IM.UBALFUL)
        W(U5)=-W(U1)*(PURC,PROFIT,KWH) $ COSTS($/KWH*KWH/D)
        W(U6)=-W(U2)*(PURC,PROFIT,H2O)$COST$/MGAL* MGAL/DAY
        LINE,T8='POWER, KWH/SD',E(VOL)39=W(U1),E(CST)55=W(U5)
        LINE,T8='COOLING WATER, MGAL/SD',E(VOL)39=W(U2),E(CST)55=W(U6)
        LINE,T8='STEAM MSCF/SD',E(VOL)39=W(U3)
        LINE,T8='FUEL, MMBTU/SD',   E(VOL)39=W(U4)
        SKIP2
LOG,PAGE
```

Table 10-5 Refinery cost model

REFINERY COST MODEL

	CRUDE YIELDS	
CHARGE STREAMS	BBL/SDAY	VOL PCT
BOLIVIAN CRUDE	26,316	79.7
COLUMBIAN CRUDE	6,706	20.3
	-------	------
TOTAL	33,022	100.0
YIELD STREAMS		
COLNP COLUMBIA NAP	1,966	6.0
BOLNP BOLIVIAN NAP	14,131	42.8
KEROP KEROSINE	4,232	12.8
DIESLP SR DIESEL	3,474	10.5
BBTMP BOL VAC BTM	974	2.9
CBTMP COL VAC BTM	1,207	3.7
AGOP ATM BOT	1,788	5.4
VACGOP VAC GAS OIL	5,250	15.9
	-------	------
TOTAL	33,022	100.0
UTILITY SUMMARY	AMOUNT	COST,$/SD
POWER,KWH/SD	10,522	420.88
COOLING WATER,MGAL/SD	5,318	280.79
STEAM MSCF/SD	99	
FUEL, MMBTU/SD	2,251	

certain portions of the MPS solution file and print them according to a specified layout. They are the other facet to matrix generator languages, discussed in Sec. 8-4.

To complete the presentation of our example problem, the instructions (in GAMMA) for preparation of a report are shown in Table 10-4, and the ensuing report is shown in Table 10-5.

To round out our discussion of the example problem, we will present the results of parametric variation of the FCCU capacity (and the alkylation plant, which is assumed to vary in direct proportion). The results shown in Table 10-6 are for hydrocracker capacity of 3.870 MBSD (incurring an additional investment of 4 $MM, see Sec. 7-2) and crude unit overhead capacity of 23.26 MBSD (incurring an additional investment of 1.2 $MM). The total investment cost listed in column 2 of Table 10-6 includes these sums.

Note that we are effectively performing a single-variable optimization on the FCCU capacity, whose economies of scale make it difficult to include directly in the objective function. Rather than perform the calculations externally, special ordered sets, discussed in Sec. 6-4-3, may be used in conjunction with a piecewise linear approximation to the six-tenths power law relating investment cost to capacity.

On the basis of maximum incremental nett operating profit as listed in column 5 of Table 10-6, the principal recommendation to management would be to construct a 14.74 MBSD FCCU. Other supplementary questions could be explored and reported (indeed, this is the main reason for having an LP model); for example, the fairly high shadow prices of hydrocracker capacity (row VCAPHOL) and vacuum unit overhead capacity (row VCAPHVO) suggest they should be increased.

Table 10-6 Results of parametric variation of FCCU capacity

FCCU capacity (MBSD)	Total investment (MM$)	Amortized investment cost (MM$/yr) $n = 10$, $i = 10\%$	Incremental operating profit[1] (MM$/yr)	Incremental nett profit[2] (Col. 4 – Col. 3) (MM$/yr)
0	5.2	0.85	5.14	4.29
2	21.53	3.50	11.62	8.12
4	29.96	4.87	17.43	12.56
6	36.77	5.98	20.33	14.35
7.26	40.60	6.61	22.02	15.41
8	42.72	6.95	22.90	15.95
10	48.10	7.83	25.19	17.36
12	53.05	8.63	27.48	18.85
14	57.69	9.39	29.78	20.39
14.74[3]	59.35	9.66	30.58	20.92
16	62.07	10.10	30.58	20.48

(1) Incremental operating profit is operating profit (LP solution) minus operating profit for base case (refinery as is) which is 21.72 MM$/yr on 347 stream-days per year (95 percent on-stream factor).

(2) Incremental nett profit before labor, maintenance, taxes, insurance, depreciation, sales expense, and corporate overhead.

(3) Point at which capacity ceases to be fully utilized.

A final cautionary note should also be made that answers to investment decisions like our example problem depend very much on the criteria used; for example, using the internal rate of return† as a criterion would suggest having no FCCU, since a minimal amount of investment is required to realize the extra capacity of the hydrocracker and crude unit overhead streams; thus the ratio column 2/column 4 in Table 10-6 is very much smaller for this case.

The conclusion to be drawn is that the early gains are the easiest to make, and the investment decision depends very much upon the available budget.

† The internal rate of return can be defined for our example problem as the interest rate i, given n, for which

$$(\text{col. } 2) \times \frac{i(1+i)^n}{(1+i)^n - 1} = (\text{col. } 4) \text{ in Table 10.1}$$

APPENDIX

A

PDS/MAGEN PROGRAM

As a comparison to the GAMMA language used in the text, the following PDS/MAGEN program to generate the MPS file of our example problem from the data tables was kindly provided by John Green of Control Data Corporation.

It should be emphasized that the layout of the tables in Sec. 8-2 made programming fairly straightforward; both GAMMA and PDS/MAGEN are considerably more powerful than they were required to be for this example. In particular, they can perform arithmetic operations to construct matrix entries from raw data.

In PDS/MAGEN, it is necessary to have the descriptive text appearing first in the data tables, so a typical table has the format shown in Table A-1. Note that the symbols ≡ or # indicate the end of description and the symbol + indicates a continuation of the table. As in GAMMA, the symbol * indicates a comment.

The definition of lists occurs in the DICTIONARY section, where CLASSES are defined in an analogous fashion to GAMMA (see Table A-2). Note, as in GAMMA, that brackets distinguish a list, or CLASS, from an individual element.

The table RTYPE in Table A-3 indicates the row types. This is followed by a series of statements which have the effect of creating a table ROW, containing all the rownames and a dummy column T with entries of 0.0. We then place in column T the appropriate row types indicated in table RTYPE.

The command FOR ROWNAME = TABLE (TABL) (,1) creates a temporary list (ROWNAME) consisting of all rownames in the list of tables (TABL), starting at the first character. As in GAMMA, the phrase (ROWN/ROWNAME) accesses the list (ROWN) while indexing on the list (ROWNAME) (cf. Sec. 8-4 example).

The ROWS section of the MPS file is created from the information in table ROW (see Table A-4). The command COPY copies the subsequent text directly to the MPS file, e.g., NAME, ROWS, COLUMNS.

The COLUMNS section is formed in the same manner as in GAMMA, i.e., using lists for table, row, and column entries. Note that the order of access is (Table, Column, Row). PDS/MAGEN will suppress generation of a detail line when a referenced table entry is empty, unless otherwise directed.

The RHS is generated by forming the column vector, called LIMTMAX, from the entries in the table LIMT, column MAX, row (list) (ROWNAME).

PDS/MAGEN also has report-writing capabilities similar to GAMMA, and we may use code similar to Table 10-4 to generate a report in the form of Table 10-5.

Table A-1 Data table for PDS/MAGEN

```
        TABLE VSGP          SATURATE GAS PLT
                                  *      LNC         LNB         LHG
      COLNPG COL NAPH          MVOLLNC   1
      BOLNPG BOL NAPH          MVOLLNB               1
      95BBG REFORMER95         MVOLB95
      90BBG REFORMER 90        MVOLB90
      HYDBBG UNICRACKER        MVOLLHG                           1
      C3P PROPANE              MVOLC3S   -.0112      -.0277      -.175
      C4P BUTANE               MVOLNC4   -.0378      -.0563      -.270
      LSRNAP  LSR NAPH         MVOLLSR   -.1502      -.1990      -.028
      HSRNAP HSR NAPH          MVOLHSR   -.7953      -.6873
      IC4P ISO BUTANE          MVOLIC4   -.0099      -.017       -.455
      KWH                      UBALKWH   .721        .495        .495
      H2O                      UBALH2O   .185         .209        .209
      STM                      UBALSTM   .013        .013        .013
      FUEL                     UBALFUL   .0488       .0506       .0448
      CAP SAT GAS PLT          VCAPSGP   1           1
      COST                     PROFIT    0           0           0
*
*                         SATURATE GAS PLANT - CONTINUED
*
+                                        B95         B90
                               MVOLB95   1
                               MVOLB90               1
                               MVOLC3S   -.2836      -.2710
                               MVOLNC4   -.3285      -.3289
                               MVOLLSR   -.0241      -.0255
                               MVOLHSR
                               MVOLIC4   -.2502      -.2656
                               UBALKWH   .495        .495
                               UBALH2O   .209        .209
                               UBALSTM   .0130       .0130
                               UBALFUL   .0506       .0506
                               VCAPSGP
                               PROFIT    0           0
*
```

Table A-2 Dictionary section of PDS/MAGEN

```
DICTIONARY
    CLASS UNIT
          VCRD  CRUDE UNIT
          VSGP  SATURATE GAS PLANT
          VH2R  UNIFINER
          VRFF  PLATFORMER FEED
          VRFG  PLATFORMER
          VHOL  UNICRACKER
          VCCU  FLUID CAT CRACKER
          VALK  ALKYLATION PLANT
          UTIL
          PURC
    CLASS SLS
          LPG  LPG
          JP4  JP4
          DSL  DIESEL
          PRE GASOLINES: PREMIUM
          INT            MEDIUM
          REG            REGULAR
          RSD  RESIDUUM
    CLASS BLND
          B(SLS)
    CLASS TABL
          (UNIT)
          (BLND)
```

Table A-3 Preparation of row types

```
*
*
         TABLE RTYPE        TABLE OF ROW TYPES
                    T
      MVOL         'L'
      EVOL         'E'
      VCAP         'L'
      MSCF         'E'
      UBAL         'E'
      PROF         'N'
      X            'L'
*  CREATE TABLE OF ALL ROWNAMES
     FORM SECTION (TABL)
     FORM SECTION (ROWNAME)  FOR ROWNAME = TABLE (TABL)(,1)
     FORM TABLE ROW(T,(ROWNAME)) = 0
     FORM SECTION,END (ROWNAME)
     FORM SECTION,END (TABL)
*
*   CODE ROW TYPE
     FORM SECTION (ROWN/ROWNAME)
                    FOR ROWNAME = TABLE ROW(,1)
                    FOR  ROWN = TABLE ROW(,1)
     FORM TABLE ROW(T,(ROWNAME)) = TABLE RTYPE(T,(ROWN/ROWNAME))
     FORM TABLE ROW(T,(ROWNAME)) = TABLE RTYPE(T,(R/ROWNAME))
                    FOR R= TABLE ROW(,1)
     FORM SECTION,END (ROWN/ROWNAME)
*
```

Table A-4 MPS file generation

```
*         SET UP ROWS SECTION
     COPY
NAME            OIL REFINERY EXAMPLE
ROWS
     FORM SECTION (ROWNAME)
                               FOR ROWNAME = TABLE ROW(,1)
     FORM SECTION (T)
                               FOR T = TABLE ROW(T,(ROWNAME))
     FORM ROW ID
         (ROWNAME) = (T)
     FORM SECTION,END(T)
     FORM SECTION,END (ROWNAME)
*
*
*          SET UP COLUMNS SECTION
     COPY
COLUMNS
     FORM SECTION (UNIT)
     FORM SECTION (COL)
                               FOR COL = TABLE (UNIT)(1,)
     FORM VECTOR (UNIT)(COL)
         (ROWNAME) = TABLE (UNIT)((COL),(ROWNAME))
                               FOR ROWNAME = TABLE (UNIT)(,1)
     FORM SECTION,END (COL)
     FORM SECTION,END (UNIT)
     FORM SECTION (SLS)
     FORM VECTOR B(SLS)(COL)
                               FOR COL = TABLE B(SLS)(1,)
                               EXCEPT COL = (SLS)
         (ROWNAME) = TABLE B(SLS)((COL),(ROWNAME))
                               FOR ROWNAME = TABLE B(SLS)(,1)
     FORM VECTOR SELL(COL)
                               FOR COL = TABLE B(SLS)(1,)
                               WHEN COL = (SLS)
         (ROWNAME) = TABLE B(SLS)((COL),(ROWNAME))
                               FOR ROWNAME = TABLE B(SLS)(,1)
     FORM SECTION,END (SLS)
*
*
*           SET UP RHS SECTION
     COPY
RHS
     FORM VECTOR LIMTMAX
         (ROWNAME) = TABLE LIMT(MAX,(ROWNAME))
                               FOR ROWNAME = TABLE LIMT(,1)
     COPY
ENDATA
```

BIBLIOGRAPHY

APEX III (1977) *User Manual, Control Data Corporation*, Minneapolis.

Bartels, R. H., and G. H. Golub (1969) "The Simplex method of linear programming using LU decomposition," *Communication of ACM*, **12**: 266–268.

Beale, E. M. L. (1967) "Numerical methods," in J. Abadie (ed.), *Nonlinear Programming*, North-Holland, Amsterdam, pp. 132–205.

——— (1968) *Mathematical Programming in Practice*, Pitmans, London.

——— (1970) "Advanced algorithmic features for general mathematical programming systems," in J. Abadie (ed.), *Integer and Nonlinear Programming*, North-Holland, Amsterdam, pp. 119–137.

——— (1971) "Sparseness in linear programming," in J. K. Reid (ed.), *Large Sparse Sets of Linear Equations*, Academic Press, London, pp. 1–16.

——— (1976) "Optimization techniques based on linear programming," in L. C. W. Dixon (ed.), *Optimization in Action*, Academic Press, London, pp. 447–466.

——— and J. A. Tomlin (1970) "Special facilities in a general mathematical programming system for non-convex problems using ordered sets of variables," in J. Lawrence (ed.), *Proceedings of the Fifth International Conference on Operational Research*, Tavistock, London, pp. 447–454.

Bodewig, E. (1959) *Matrix Calculus*, North-Holland, Amsterdam.

Brearley, A. L., G. Mitra, and H. P. Williams (1975) "Analysis of mathematical programming problems prior to applying the simplex method," *Mathematical Programming*, **8**: 54–83.

Buzby, B. R. (1974) "Techniques and experience solving really big nonlinear programmes," in R. Cottle and J. Krarup (eds), *Optimization Methods for Resource Allocation*, English Universities Press, London, pp. 229–236.

Dahlquist, G., and A. Bjork (1974) *Numerical Methods*, Prentice-Hall, New Jersey.

Dantzig, G. B. (1963) *Linear Programming and Extensions*, Princeton University Press, New Jersey.

———, R. P. Harvey, R. D. McKnight, and S. T. Smith (1969) "Sparse matrix techniques in two mathematical programming codes," in R. A. Willoughby (ed.), *Sparse Matrix Proceedings*, Report RA1, IBM Corporation, New York.

——— and R. M. Van Slyke (1967) "Generalized upper bounding techniques," *Journal of Computer and System Science*, **1**: 213–216.

——— and P. Wolfe (1960) "Decomposition principle for linear programmes," *Operations Research*, **8**: 101–110.

de Buchet, J. (1971) "How to take into account the low density of matrices to design a mathematical programming package," in J. K. Reid (ed.), *Large Sparse Sets of Linear Equations*, Academic Press, London, pp. 211–218.

Driebeek, N. J. (1966) "An algorithm for the solution of mixed-integer programming problems," *Management Science*, **12**: 576–587.

Duff, I. S. (1976) "A survey of sparse matrix research," AERE Report C.S.S. 28, Harwell, England.

Fieldhouse, M. (1974) "Checking large scale LP models," in R. Cottle and J. Krarup (eds), *Optimization Methods for Resource Allocation*, English Universities Press, London, pp. 165–175.

Forrest, J. J. H., and J. A. Tomlin (1972) "Updating triangular factors of the basis to maintain sparsity in the product-form simplex method," *Mathematical Programming*, **2**: 263–278.

———, J. P. H. Hirst, and J. A. Tomlin (1974) "Practical solution of large and complex integer programming problems with UMPIRE," *Management Science*, **20**: 736–773.

Gal, T. (1979) *Postoptimal Analyses, Parametric Programming and Related Topics*, McGraw-Hill, New York.

Garfinkel, R. S., and G. L. Nemhauser (1972) *Integer Programming*, Wiley, New York.

Gill, P. E., G. H. Golub, W. Murray, and M. A. Saunders (1974) "Methods for modifying matrix factorizations," *Mathematics of Computation*, **28**: 505–535.

———, and W. Murray (1973) "A numerically stable form of the simplex algorithm," *J. Linear Algebra Applics.*, **7**: 99–138.

———, ———, and M. A. Saunders (1975) "Methods for computing and modifying the LDV factors of a matrix," *Mathematics of Computation*, **29**: 1051–1077.

Goldfarb, D., and J. K. Reid (1977) "A practicable steepest-edge simplex algorithm," *Mathematical Programming*, **12**: 361–371.

Griffith, R. E., and R. A. Stewart (1961) "A nonlinear programming technique for the optimization of continuous processing systems," *Management Science*, **7**: 379–392.

Hamming, R. W. (1971) *Introduction to Applied Numerical Analysis*, McGraw-Hill, New York.

Harris, P. M. J. (1973) "Pivot selection methods of the Devex LP code," *Mathematical Programming*, **5**: 1–28.

Hellerman, E., and D. C. Rarick (1971) "Reinversion with the preassigned pivot procedure," *Mathematical Programming*, **1**: 195–216.

——— and ——— (1972) "The partitioned preassigned pivot procedure," in D. J. Rose and R. A. Willoughby (eds), *Sparse Matrices and their Applications*, Plenum Press, New York, pp. 68–76.

Kalan, J. E. (1971) "Aspects of large-scale in-core linear programming," *Proceedings of ACM Conference*, Chicago, 304–313.

Markowitz, H. M. (1957) "The elimination form of the inverse and its application to linear programming," *Management Science*, **3**: 255–269.

Miller, C. E. (1963) "The simplex method for local separable programming," in R. L. Graves and P. Wolfe (eds), *Recent Advances in Mathematical Programming*, McGraw-Hill, New York, pp. 89–100.

Murtagh, B. A., and M. A. Saunders (1978) "Large scale linearly constrained optimization," *Mathematical Programming*, **14**: 41–72.

——— and ——— (1980) "The implementation of a Lagrangian-based algorithm for sparse nonlinear constraints," Systems Optimization Laboratory Report SOL 80–1, Stanford University, California.

Orchard-Hays, W. (1974) "On the proper use of a powerful MPS," in R. Cottle and J. Krarup (eds), *Optimization Methods for Resource Allocation*, English Universities Press, London, 229–236.

Rarick, D. C. (1975) "An improved pivot row selection procedure, implemented in the mathematical programming system MPS III," *Management Science Systems*, Rockville, Maryland.

Robinson, S. M. (1972) "A quadratically convergent algorithm for general programming problems," *Mathematical Programming*, **3**: 145–156.

Salkin, G. and J. Kornbluth (1973) *Linear Programming in Financial Planning*, Accountancy Age, London.

Sargent, R. W. H., and B. A. Murtagh (1973) "Projection methods for nonlinear programming," *Mathematical Programming*, **4**: 245–268.

Saunders, M. A. (1972) "Large-scale linear programming using the Cholesky factorization," Report STAN-CS-72-252, Stanford University, California.

——— (1976) "A fast, stable implementation of the simplex method using Bartels-Golub updating," in J. R. Bunch and D. J. Rose (eds), *Sparse Matrix Computations*, Academic Press, New York and London, pp. 213–226.

Sherman, J., and W. J. Morrison (1949) "Adjustment of an inverse matrix corresponding to changes in the elements of a given column or a given row of the original matrix," *Ann. Math. Stat.*, **20**: 621.

Simonnard, M. (1966) *Linear Programming*, translated by W. S. Jewell, Prentice-Hall, New Jersey.

Smith, B. R., P. D. Lucas, and B. A. Murtagh (1976a) "The development of a New Zealand energy model," *N.Z. Operational Research*, **4**: 101–117.

———, ———, and ——— (1976b) "Some aspects of modelling New Zealand's electricity sector," *N.Z. Energy Journal*, **49**: 100–103.

Stephenson, G. G. (1970) "A hierarchy of models for planning in a division of ICI," *Operational Research Quarterly*, **21**(2): 221–245.

Stewart, G. W. (1973) *Introduction to Matrix Computations*, Academic Press, New York.

Strang, G. (1976) *Linear Algebra and Its Applications*, Academic Press, New York.

Tomlin, J. A. (1970) "Branch and bound methods for integer and non-convex programming," in J. Abadie (ed.), *Integer and Nonlinear Programming*, North-Holland, Amsterdam, pp. 437–450.

Williams, H. P. (1978) *Model Building in Mathematical Programming*, Wiley, Chichester.

Wolfe, P. (1969) "Trends in linear programming computation," in R. A. Willoughby (ed.), *Sparse Matrix Proceedings*, Report RA1, IBM Corporation, New York.

INDEX